Student Solutions Manual

for use with

Prepared by KEN SILLS
and
STEVE SHORLIN

NELSON EDUCATION

NELSON EDUCATION

Student Solutions Manual for use with Physics for the Life Sciences
by Ken Sills and Steve Shorlin

Associate Vice President, Editorial Director:
Evelyn Veitch

Editor-in-Chief, Higher Education:
Anne Williams

Executive Editor:
Paul Fam

Marketing Manager:
Sean Chamberland

Developmental Editor:
Tracy Yan

Manufacturing Coordinator:
Pauline Long

Printer:
Webcom

Printed and bound in Canada
2 3 4 11 10 09 08

For more information contact Nelson Education Ltd.
1120 Birchmount Road,
Toronto, Ontario, M1K 5G4.
Or you can visit our internet site at http://www.nelson.com

ISBN-13: 978-0-17-647276-4
ISBN-10: 0-17-647276-2

TABLE OF CONTENTS

PREFACE

We are excited to provide students with a set of detailed solutions corresponding to the odd-numbered concept questions and problems found at the end of each chapter in Nelson Education's *Physics for the Life Sciences*. We believe that this material supplements the textbook, and provides the best possible preparation in physics for those interested in pursuing careers and further studies in the life sciences.

To meet the various objectives of a solutions manual with a modern first-year textbook, we chose to mix two different styles:

1. Half of the problem-solving solutions are written in a tutorial style. This allows students to work on the material independently, either during self-study, to deepen the understanding of the concepts presented in the textbook, or as an accompaniment to assigned homework, confirming their approach without having to wait for the next tutorial session.

2. The other half of the problem-solving solutions includes extended hints or brief synopses of the solutions. These solutions are intended to encourage study in groups, either with a tutorial-session leader or in a peer-driven atmosphere.

The solutions manual provides guidance and correct answers to check with the group's results. As well, the answers to the open-ended concept questions have been designed to stimulate discussion among students. We hope that students will be allowed the time necessary to contemplate and experience the enjoyment that comes from thinking about real-world problems.

The primary purpose of this textbook and its solutions manual is to engage students of the life sciences with the idea that physics is just beneath the surface of any problem worth studying. The contents, as well as the methods used of presentation, have been guided by state-of-the-art physics education research, as well as years of experience teaching undergraduate physics at universities with excellent medical schools.

The authors wish to acknowledge the support of those who have helped with the preparation of this manuscript: Paul Fam, Acquisitions Editor, and Tracy Yan, Developmental Editor at Nelson Education Ltd., and Mira Rasche for her extensive artwork. We are also very much indebted to Martin Zinke-Allmang for his dedication to this project.

Ken Sills
Steve Shorlin

CHAPTER ONE

Physics and the Life Sciences: An introduction

MULTIPLE CHOICE AND CONCEPTUAL QUESTIONS

Question 1.1

The statement, "We are smarter than chimpanzees because we have a bigger brain" had better be wrong, because otherwise Fig. 1.2 would imply that we are dumber than porpoises and blue whales. Based on Fig. 1.2, how would you formulate a similar comparative statement about the human brain mass that would address our perception that we are also smarter than *Cetacea* with bigger brains?

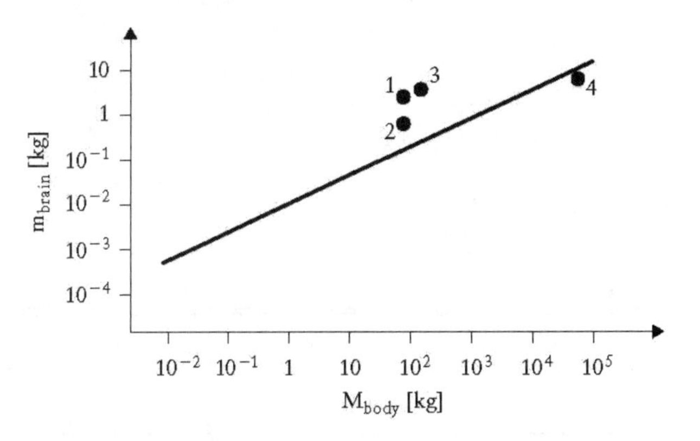

FIGURE 1.2

<u>Answer</u>: Brain size alone has no direct link to intelligence but scales simply with body size (solid line in Fig. 1.2). It could be postulated that intelligence is related to the distance above the solid line shown in Fig. 1.2. Perhaps this excess brain mass represents a larger cerebrum, which is the part of the brain associated with intelligence. This theory would suggest that porpoises are approximately as intelligent as humans.

Question 1.3

(a) If m_{brain} and M_{body} were directly proportional to each other, i.e., $m_{brain} \propto M_{body}$, the slope of the line in Fig. 1.2 would have to be (A) zero; (B) – 1; (C) + 1; (D) – 2; (E) + 2.

(b) If we replot Fig. 1.2 with the brain mass shown in unit gram (*g*), the slope (A) increases; (B) decreases; (C) remains unchanged; (D) cannot be predicted before the replotting of Fig. 1.2 is completed.

<u>Answer to part (a)</u>: (C). m_{brain} directly proportional to M_{body} means that $b = +1$ in:

$$m_{brain} = a \cdot M_{body}^{b} \qquad (1)$$

Note that the coefficient b is the slope of the curve after the natural logarithm is taken on both sides of Eq. [1].

<u>Answer to part (b)</u>: (C). We can argue in two ways: physically, we note that the slope represents an actual physical relation. Replotting data by using another unit system cannot change the physical facts. Or mathematically, plotting m_{brain} in unit *g* means that we use values that are larger by a factor of 1000 on the left side in Eq. [1]. Thus, for Eq. [1] to remain correct, the prefactor must also be larger by a factor of 1000. In Eq. [2] we take the natural logarithm on both sides of Eq. [1], with the brain mass in unit *kg* on the right–hand side:

$$\ln m_{brain}(kg) = \ln a + b \cdot \ln M_{body}(kg) \qquad (2)$$

In Eq. [3] we rewrite Eq. [1] once more with natural logarithms, but use the brain mass in unit *g*:

$$\ln m_{brain}(g) = \ln(1000\, a) + b \cdot \ln M_{body}(kg) \qquad (3)$$

in which $\ln(1000 \cdot a) = \ln 1000 + \ln a$.

Eqs. [2] and [3] differ only in that a constant term, $\ln 1000 = 6.908$, is added in the second case. This represents a vertical shift of the curve, but not a change in its slope.

ANALYTICAL PROBLEMS

Problem 1.1

(a) Plot in double–logarithmic representation the two functions (I) $y = 4x^2$ and (II) $y = 4x^2 + 1$ in the interval $0.1 \le x \le 10.0$.

(b) What draw–back of double–logarithmic plots can you identify?

<u>Solution to part (a)</u>: The plot is shown in Fig. 1.10.

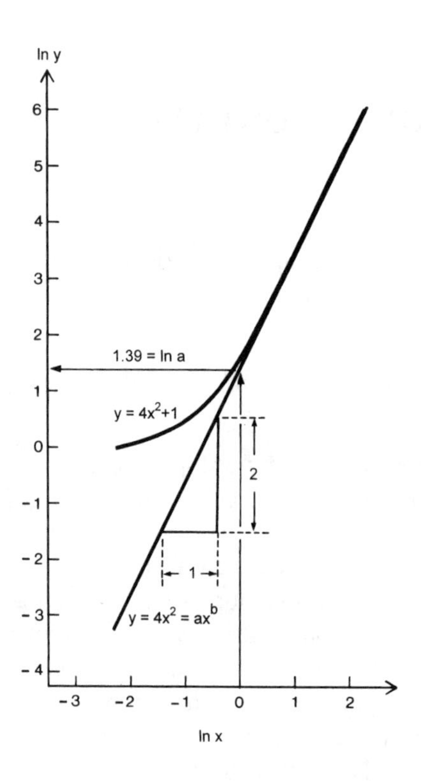

Figure 1.10

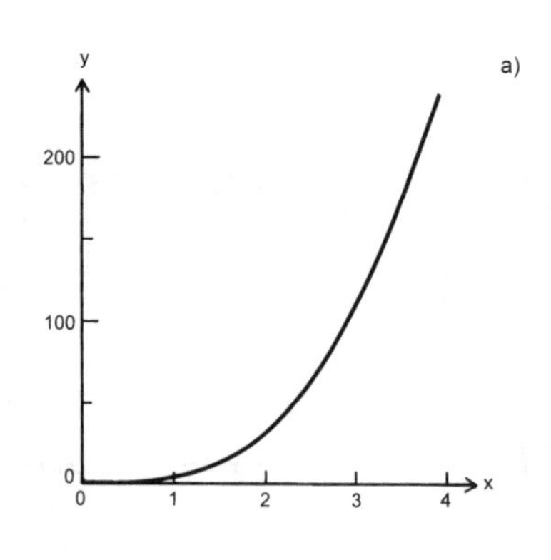

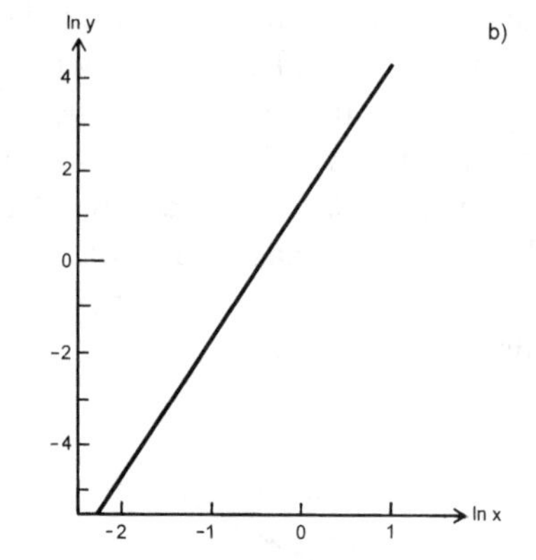

Figure 1.11

Solution to part (b): Double–logarithmic plots may be misleading when the function $y = f(x)$ is a sum. In the current case, one term dominates the double–logarithmic plot for large values of x and the other term dominates the plot for small values of x. At intermediate values the double–logarithmic plot is curved. This behaviour is apparent in Fig. 1.10.

Problem 1.3
For the function $y = 4x^3$
(a) plot y versus x for $0 \le x \le 4$,
(b) plot $\ln y$ versus $\ln x$ for $0 \le x \le 4$,
(c) show that the slope of the double–logarithmic plot is 3, and
(d) show that the intercept of the double–logarithmic plot is ln 4.

Solution to parts (a) and (b): The linear plot is shown in Fig. 1.11(a) and the double–logarithmic plot is shown in Fig. 1.11(b). Note that the range of Fig. 1.11(b) is only a subset of the range $0 \le x \le 4$, since $\ln(0) = -\infty$.

Solution to part (c): The slope in the double–logarithmic plot represents the exponent of the function, in the current problem with a value of 3. To quantify this from the plot, any data pair can be used. Such pairs at $x = 1$ and $x = 2$ are shown in Table 1.5.

Table 1.5

Data Set	$\ln x$	$\ln y$
# 1	0.0 ($x = 1.0$)	1.386
# 2	0.693 ($x = 2.0$)	3.466

In the table, the difference between the two $\ln x$ values is 0.693 and the difference between the two $\ln y$ values is $3.466 - 1.386 = 2.08$; the ratio $2.08/0.693 = 3.0$, as expected.

Solution to part (d): The intercept is read off the graph in Fig. 1.11(b) for $\ln x = 0$. At that value we find $\ln y = 1.386$ (see Table 1.5), which corresponds to $y = 4.0$.

Problem 1.5
We develop an empirical formula connecting the wingspan and the mass of some species able to fly. Then we evaluate a few interesting consequences. (The first to make these considerations was Leonardo da Vinci).
(a) Use the data in Table 1.1 to draw a double–logarithmic plot $\ln W$ versus $\ln M$ where W is the wingspan and M is the mass. Determine the constants a and b in a power–law relation $W = a \cdot M^b$.

Table 1.1

Bird	Wingspan (*cm*)	Mass (*g*)
Hummingbird	7	10
Sparrow	15	50
Dove	50	400
Andean condor	320	11500
California condor	290	12000

(b) The largest animal believed ever to fly was a late Cretaceous pterosaur species found in Texas and named *Quetzalcoatlus northropi*. It had an 11-m wingspan. What is the maximum mass of this pterosaur? *Note*: the largest wingspan of a living species is 3.6 m, for the wandering albatross.

(c) Assume that a human wishes to fly like a bird. What minimum wingspan would be needed for a person of 70 kg to take off?

<u>Solution to part (a)</u>: The resulting double–logarithmic plot is shown in Fig. 1.12. An organized approach to plotting these data is based on extending Table 1.1 to include the logarithm values of wingspan and mass, as shown in Table 1.6.

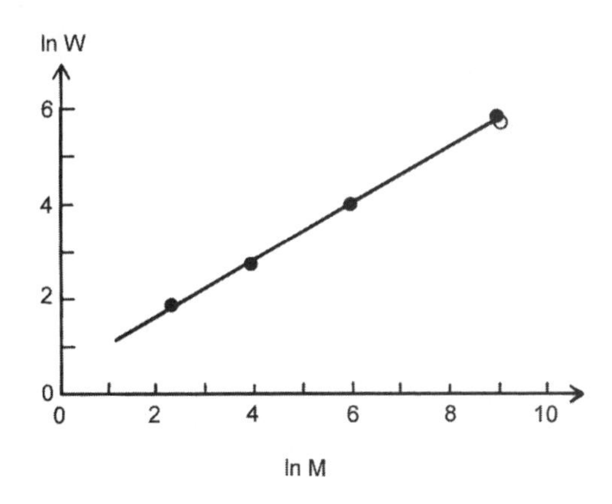

Figure 1.12

Table 1.6

Bird	W (*cm*)	ln W	M (*g*)	ln M
Hummingbird	7	1.95	10	2.30
Sparrow	15	2.71	50	3.91
Dove	50	3.91	400	5.99
Andean condor	320	5.77	11500	9.35
California condor	290	5.67	12000	9.39

Using these logarithmic data, the given power law $W = a \cdot M^b$ is rewritten in the form $\ln W = b \cdot \ln M + \ln a$. The constants a and b are determined from this equation in the manner described in the Appendix *Graph analysis methods* on p. 9 – 12 in Chapter 1. For the analysis we do not choose data pairs from Table 1.6. As the graph in Fig. 1.12 illustrates, actual data points deviate from the line that best fits the data (represented by the solid line). To avoid that the deviation of actual data affects our results, the two data pairs used in the analysis are obtained directly from the solid line in Fig. 1.12. We choose $\ln W_1 = 2$ with $\ln M_1 = 2.6$ and $\ln W_2 = 6$ with $\ln M_2 = 9.6$. This leads to:

$$\begin{array}{ll} (I) & 2.0 = \ln(a) + 2.6 \cdot b \\ (II) & 6.0 = \ln(a) + 9.6 \cdot b \\ \hline (II) - (I) & 4.0 = (9.6 - 2.6)\, b \end{array} \qquad \textbf{(4)}$$

Thus, $b = 0.57$. Due to the fluctuations of the original data and the systematic errors you commit when reading data off a given plot, values in the interval $0.5 \leq b \leq 0.6$ may have been obtained.

Substituting the value we found for b in formula (I) of Eq. [4] yields: $2 = 1.48 + \ln a$, i.e., $\ln a = 0.52$ which corresponds to a value of $a = 1.7$.

<u>Solution to part (b)</u>: We use the given value for the pterosaurs' wingspan: $W = 11$ m $= 1100$ cm. The value has been converted to unit cm since that is the unit used when we developed our formula in part (a). We first rewrite the formula for the wingspan with the mass as the dependent variable:

$$W = a \cdot M^b \quad \Rightarrow \quad M = \left(\frac{W}{a}\right)^{1/b} \qquad \textbf{(5)}$$

Entering the given value for the wingspan then leads to:

$$M = \left(\frac{1100}{1.7}\right)^{1/0.57} = 85400\ g = 85.4\ kg \qquad \textbf{(6)}$$

The mass of pterosaurs did not exceed 85 kg.

<u>Solution to part (c)</u>: We use again the power law relation we found in part (a) and insert the given value for the person's mass:

$$W = 1.7 \cdot 70000^{0.57} = 580\ cm \qquad (7)$$

A person of mass 70 kg would need a 5.8 m wing span. Notice that a pterosaur has a only 20 % larger mass than a human, but requires a 90 % increase in wingspan in order to achieve flight. Whereas a sparrow has a 500 % larger mass than a hummingbird, yet it requires only a 100 % increase in wingspan This shows the exponential nature of this relationship.

CHAPTER TWO

Locomotion I: Kinematics

MULTIPLE CHOICE AND CONCEPTUAL QUESTIONS

Question 2.1
(a) What is the sum of the two vectors **a** = (5, 5) and **b** = (–14, 5)?
(b) What are the magnitude and direction of **a** + **b**?

Answer to part (a): We define the vector sum as **r**, i.e., **r** = **a** + **b**, or, in component notation,

$$r_x = a_x + b_x = 5 + (-14) = -9$$
$$r_y = a_y + b_y = 5 + 5 = 10 \qquad \textbf{(1)}$$

Thus, **r** = (–9, 10). This vector is shown in Fig. 2.28.

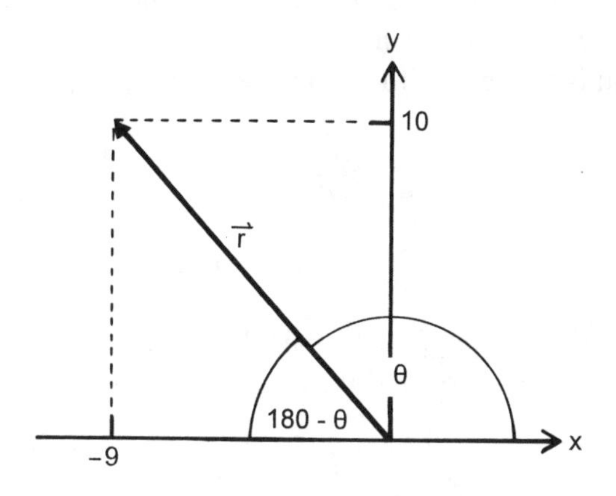

Figure 2.28

Solution to part (b): The magnitude of vector **r** is calculated with the Pythagorean theorem:

$$|r| = \sqrt{r_x^2 + r_y^2} = \sqrt{(-9)^2 + 10^2} = 13.5 \qquad \textbf{(2)}$$

To characterize the vector **r** in polar coordinates, its angle with the positive x–axis must also be determined. This can be done geometrically, as can be seen from Fig. 2.28:

$$\tan\theta = \frac{r_y}{r_x} = -1.11 \quad \Rightarrow \quad \theta = 132^0 \qquad \textbf{(3)}$$

Note: Your pocket calculator may show $\theta = -48^0$. Take care to always draw the vector to check that your answer makes sense. To your calculator, –a/b cannot be distinguished from a/(–b). It is the job of the person using the calculator to interpret the physical meaning of the number the calculator returns.

Question 2.3
We become uncomfortable if an elevator accelerates downward at a rate such that it reaches or exceeds a velocity of 6 m/s while travelling ten floors (30 metres). What value comes closest to the acceleration of such an elevator? (A) 0.3 m/s², (B) 0.6 m/s², (C) 1.2 m/s², (D) 2.0 m/s², (E) 10 m/s².

Answer: (B). Assume that the elevator starts from rest.

Question 2.5
If the average velocity of an object is zero in a given time interval, what do we know about its displacement during the same time interval?

Answer: If the average velocity in a given time interval is zero, the displacement over that time interval must also be zero.

Question 2.7
An object is thrown vertically upward.
(a) What are its velocity and acceleration when it reaches its highest altitude?
(b) What is its acceleration on its way downward half a metre above the ground?

Answer to part (a): At the moment the object reaches its highest altitude, its velocity is zero and its acceleration is equal to the acceleration due to gravity.

Answer to part (b): same as part (a) since the acceleration of the object is constant.

Question 2.9

(a) Can an object accelerate if its speed is constant?
(b) Can an object accelerate if its velocity is constant?

Answer to part (a): Yes, because the object may change its direction without changing its speed. A particular example of this will be discussed in detail when we hit uniform circular motion in Chapter 21.

Answer to part (b): No.

Question 2.11

An object is thrown upward by a person on a train that moves with constant velocity.
(a) Describe the path of the object as seen by the person throwing it.
(b) Describe the path of the object as seen by a stationary observer outside the train.

Answer to part (a): The same as if thrown on steady ground.

Answer to part (b): The path is the projectile trajectory in Eq. [2.39]. Where $v_{initial,\,x}$ is the constant horizontal velocity of the train.

ANALYTICAL PROBLEMS

Problem 2.1

A competitive sprinter needs 9.9 seconds for 100 metres. What is the average velocity in units m/s and units km/h?

Solution: $v_{average} = 10.1$ m/s = 36.4 km/h

Problem 2.3

Fig. 2.16 shows a back view of an adult male and an adult female human (accompanied by two children).

FIGURE 2.16

(a) For a typical male, the vertical distance from the bottom of the feet to the neck is $d_1 = 150$ cm and the distance from the neck to the hand is $d_2 = 80$ cm. Find the vector describing the position of the hand relative to the bottom of the feet if the angle at which the arm is held is $\theta = 35^0$ to the vertical.
(b) Repeat the calculation for a typical female with $d_1 =$ 130 cm, $d_2 = 65$ cm and the same angle θ.

Solution to part (a): We choose the origin at the bottom of the feet of the person, with the x–axis pointing horizontal and toward the right and the y–axis pointing straight up. In this coordinate system we express the two vectors shown in Fig. 2.16 for the male person, $\mathbf{d}_1$ and $\mathbf{d}_2$. Based on the given lengths (magnitudes) and the given angle we find:

$$\boldsymbol{d}_1 = \begin{pmatrix} 0 \\ |\boldsymbol{d}_1| \end{pmatrix} = \begin{pmatrix} 0 \\ 150\ cm \end{pmatrix}$$
$$\boldsymbol{d}_2 = \begin{pmatrix} |\boldsymbol{d}_2|\ \sin\theta \\ -|\boldsymbol{d}_2|\ \cos\theta \end{pmatrix} = \begin{pmatrix} 45.9\ cm \\ -65.5\ cm \end{pmatrix} \quad \textbf{(4)}$$

Note the negative sign in the y–component of the second vector! The vector from the bottom of the feet to the hand is the sum of the two vectors in Eq. [4]:

$$\boldsymbol{d}_{male} = \boldsymbol{d}_1 + \boldsymbol{d}_2 = \begin{pmatrix} 45.9\ cm \\ 84.5\ cm \end{pmatrix} \quad \textbf{(5)}$$

Solution to part (b): The calculations are analogous for the female person and yield:

$$\boldsymbol{d}_{female} = \begin{pmatrix} -37.3\ cm \\ 76.8\ cm \end{pmatrix} \quad \textbf{(6)}$$

Problem 2.5

A bacterium moves with a speed of 3.5 μm/s across a petri dish with radius $r = 8.4$ cm. How long does it take the bacterium to traverse the petri dish along its diameter?

Solution: Assume that the bacterium moves along a straight line:

$$t = \frac{d}{v} = \frac{2\ (8.4\ cm)}{3.5\ \mu m/s} = 4.8 \times 10^4\ s = 13.3\ hr \quad \textbf{(7)}$$

Actual bacteria move in random fashion; this type of motion is discussed in Chapter 10.

Problem 2.7
An object is released at time $t = 0$ upward with initial speed 5.0 m/s. Draw an $x(t)$ plot for the time period until it returns back to its initial position.

Solution:
The graph is shown in Fig. 2.29.

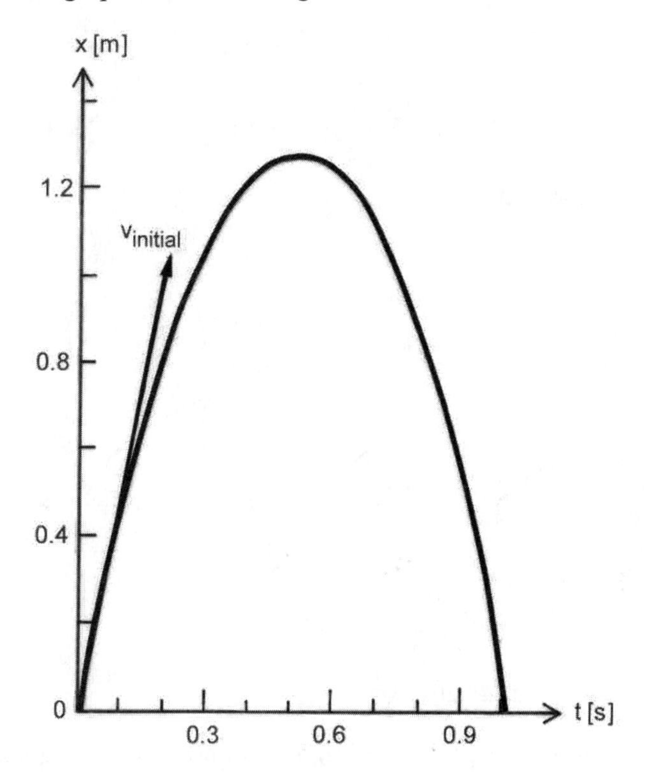

Figure 2.29

Problem 2.9
An object is dropped from rest from a height of 10 m. What is its constant acceleration upward if it hits the ground with a speed of 1 m/s?

Solution: We use $v_{\text{initial}} = 0$ in Eq. [2.25]:

$$a = g - \frac{v_{final}^2}{2 \cdot \Delta h} \tag{8}$$

$$= 9.8 \frac{m}{s^2} - \frac{(1.0\ m/s)^2}{2\ (10\ m)} = 9.75 \frac{m}{s^2}$$

Problem 2.11
The best major league baseball pitchers can throw a baseball with velocities exceeding 150 km/h. If a pitch is thrown horizontally with that velocity, how far does the ball fall vertically by the time it reaches the catcher's glove 20 metres away?

Solution: Time for baseball to reach the catcher:

$$t = \frac{\Delta x}{v_{initial,\,x}} = \frac{20\ m}{150\ km/h} = 0.48\ s \tag{9}$$

During that time the ball falls a distance:

$$y = \frac{1}{2} g \cdot t^2 = \frac{9.8\ m/s^2}{2} (0.48\ s)^2 = 1.1\ m \tag{10}$$

Problem 2.13
Fish use various techniques to escape a predator. Forty species of flying fish exist — such as the California flying fish, which has a length of 50 cm. These animals escape by leaving the water through the surface, propelled by their tails to typical speeds of 30 km/h. If the flying fish could not glide,
(a) how far would they fly through air if they left at 45^0?
(b) They can travel up to 180 metres before re–entering the water. Did they use their wing–like pectoral fins to glide?

Solution to part (a): Time for fish to reach highest point (with $v = 0$ in Eq. [2.16]):

$$t = \frac{v_{initial,\,y}}{g} = 0.6\ s \tag{11}$$

Distance travelled during time $2 \cdot t$ (until re–entry):

$$\Delta x = v_{initial,\,x} \cdot (2\ t) = 7.1\ m \tag{12}$$

Solution to part (b): Yes, the fish is definitively gliding.

CHAPTER THREE

Biomechanics: Forces and Newton's Laws

MULTIPLE CHOICE AND CONCEPTUAL QUESTIONS

Question 3.1

Fig. 3.42 shows five experimental arrangements. In part (A), the object is vertically attached to a string, in parts (B) and (C) the object is in a bowl–shaped structure, in part (D) it lays on a horizontal table and in part (E) the object is held by a string on an inclined surface. In which case is the object not in mechanical equilibrium?

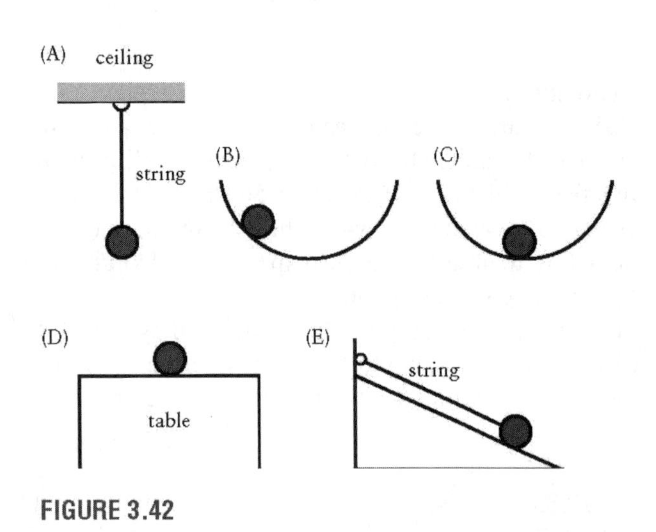

FIGURE 3.42

Answer: (B). Note that (D) is metastable if the surface is frictionless: the object in (D) lacks a force that would act to keep it in place, but also has no reason to begin moving.

Question 3.3

In Example 2.4 we discussed the Mantis shrimp's ability to achieve astonishing accelerations with its spear like arms, allowing it to kill its prey without the victim ever noticing the attack. Assume that the mantis shrimp uses a constant force to accelerate its weapon and is at a distance d_0 from its prey.

(a) Which of the three plots in Fig. 3.43 shows the acceleration as a function of position d?

(b) Which of the three plots in Fig. 3.44 shows the velocity of the weapon as a function of position d?

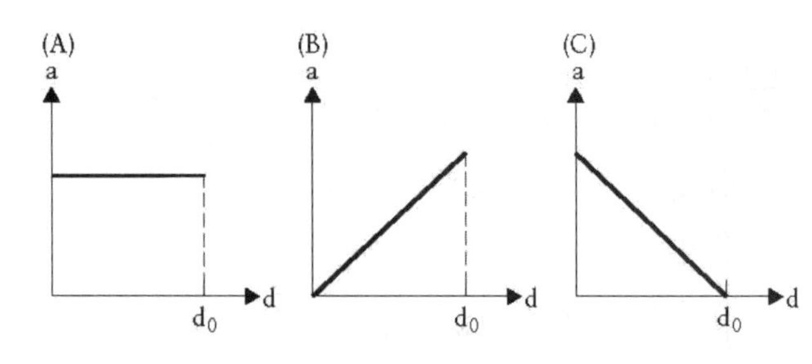

FIGURE 3.43

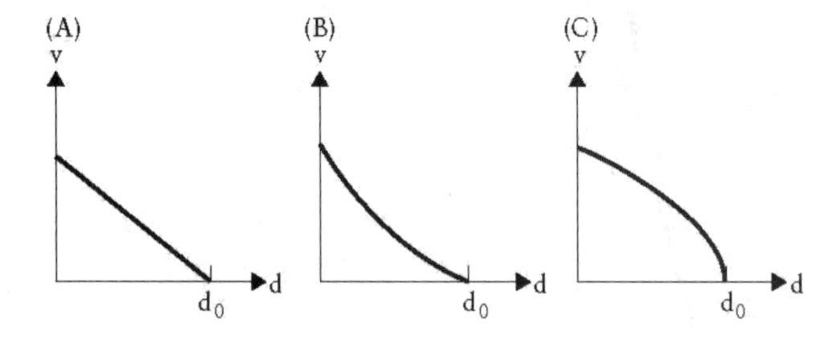

FIGURE 3.44

Answer for part (a): (A)

Answer for part (b): None of the graphs correctly depict the velocity as a function of distance. The correct graph would show that for a constant acceleration, the velocity is related to the square root of the distance travelled.

Question 3.5

The dumbbell in Fig. 3.7 has a weight $\mathbf{W}_{\mathbf{dumbbell}}$. Why do we not include this force when discussing the mechanics of the lower arm of the standard man in the figure? (A) The force $\mathbf{W}_{\mathbf{dumbbell}}$ acts in another direction than the listed forces and therefore has to be omitted; (B) The force $\mathbf{W}_{\mathbf{dumbbell}}$ has no reaction force in the figure, and for that reason cannot be considered in Newton's laws; (C) The force $\mathbf{W}_{\mathbf{dumbbell}}$ does not act on the standard man's arm and has to be excluded when Newton's law is applied to the arm; (D) We have already included the weight of the arm, $\mathbf{W}_{\mathbf{arm}}$, and no way exists to include two different weights in Newton's laws; (E) The dumbbell is not alive and can therefore not exert a force on another object.

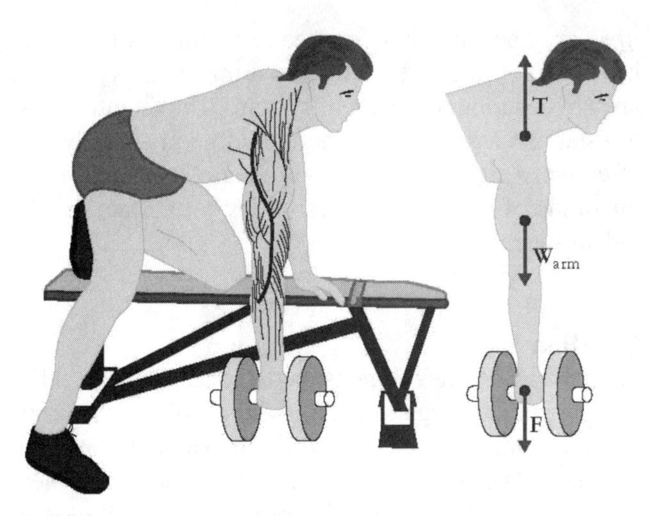

FIGURE 3.7

Answer: (C)

Question 3.7

Fig. 3.45 shows a round object on a table touching a block. Which of the following six equations is the proper application of Newton's laws in the vertical direction describing the forces acting on the round object? *Hint*: we label N the normal force, W the weight and F the contact force with the block.

$$(A)\ F + W = 0 \quad (B)\ F - W = 0$$
$$(C)\ N + W = 0 \quad (D)\ N - W = 0 \quad \mathbf{(1)}$$
$$(E)\ N - F = 0 \quad (F)\ N + F = 0$$

FIGURE 3.45

Answer: (D)

Question 3.9

Two forces act on a system, $\mathbf{F}_1 = (F_1, 0)$ along the x–axis and $\mathbf{F}_2 = (0, F_2)$ along the y–axis. What is the minimum number of forces required in this case to establish mechanical equilibrium? (A) two; (B) three; (C) four; (D) five; (E) the number depends on further information not provided.

Answer: (B). Only one additional force is needed, i.e., $\mathbf{F}_3 = (-F_1, -F_2)$. That one, plus the original two forces, makes for a total of three.

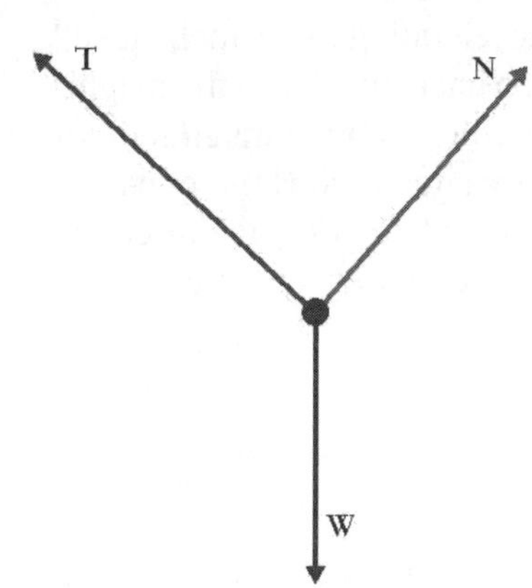

FIGURE 3.48

Question 3.11

Fig. 3.48 shows a free–body diagram with three forces, a tension **T**, a normal force **N**, and a weight **W**. For which of the five cases shown in Fig. 3.15 is this free–body diagram correct?

Answer: (A) because the angle between **T** and **N** is 90^0 and the tension is acting to the upper left.

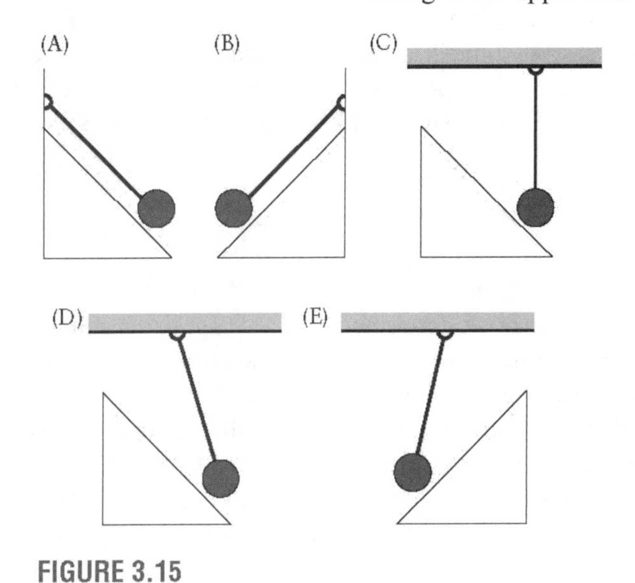

FIGURE 3.15

Question 3.13

Is it possible for an object to move if no net force acts on it?

Answer: Yes. It can maintain a constant velocity.

Question 3.15

Two forces $\mathbf{F}_1$ and $\mathbf{F}_2$ act on an object. It accelerates in a particular direction. Under what circumstances can you predict the direction of force $\mathbf{F}_1$ regardless of the specific value of force $\mathbf{F}_2$?

Answer: If $\mathbf{F}_2$ and the acceleration of the object are anti–parallel.

Question 3.17

An object of mass m accelerates with acceleration a. How does the acceleration change if we double the mass of the object but keep the accelerating force unchanged? (A) the acceleration remains unchanged; (B) the magnitude of the acceleration doubles; (C) the magnitude of the acceleration increases by a factor 4; (D) the magnitude of the acceleration is halved; (E) only the direction of the acceleration changes, but not its magnitude.

Answer: (D)

Question 3.19
Due to Newton's third law we can make the following statement about the forces in Fig. 3.7: (A) The arm exerts on the trunk a force that equals $-\mathbf{T}$; (B) The weight $\mathbf{W}_{arm}$ is equal in magnitude to the tension T; (C) Forces $\mathbf{W}_{arm}$ and $\mathbf{T}$ are an action–reaction pair of forces; (D) Forces $\mathbf{W}_{arm}$ and $\mathbf{F}$ are an action–reaction pair of forces; (E) The weight of the arm has no reaction force.

Answer: (A)

Question 3.21
Two objects are connected with a string. In the quantitative treatment of the problem we assume that the string exerts a tension force of magnitude T on each of the two objects. What has to be the case for this assumption to be valid? (A) the string hangs loose between the objects; (B) the string does not pass over a pulley; (C) the string is massless; (D) the two objects are not on a collision course with each other.

Answer: (C). Though it is also true that the tension may not exert the same magnitude of force on both objects if the string passes over a pulley that is either massive, or not frictionless.

Question 3.23
Newton reported the following observation in *Principia*: Let's assume that two objects attract each other. Contrary to the third law of mechanics, however, we assume that object B is attracted by object A more strongly than object A is attracted by object B. To test this case further, we connect objects A and B with a stiff, but massless string, forcing both objects to maintain a fixed distance. Since the force on object B is stronger than on object A, a net force acts on the combined object: it will accelerate without the action of an external force. Since this contradicts the first law of mechanics, we conclude that no violation of the third law is possible and that the third law is already included when we introduced the first law. Indeed, is the third law based on this argument not required as a separate law?

Answer: The third law is already included in the first law. Newton did not try to find a minimum system of laws to describe mechanical systems. His intention was to establish a set of laws that could be practically and effectively applied. Given that we are still using them hundreds of years later is testimony to the fact that he did so successfully.

Question 3.25
We live in a three–dimensional space. Consequently, each vestibular organ contains three orthogonal semicircular canals to measure the independent Cartesian components of the acceleration of the head. Why, then, do we have only two maculae per ear and not three?

Answer: The maculae measure each the direction of gravity relative to their surface normal. One measurement for the tilt in the xz–plane (back and forth) and one in the yz–plane (left and right) is sufficient.

ANALYTICAL PROBLEMS

Problem 3.1
One cubic centimetre ($1\ cm^3$) of water has a mass of one gram (1 g).
(a) Determine the mass of one cubic metre ($1\ m^3$) of water.
(b) Assume that a spherical bacterium consists of 98 % water and has a diameter of 1.0 μm. Calculate the mass of its water content.
(c) Modelling a fly as a water cylinder of 4 mm length and 1 mm radius, what is its mass? *Note*: Important formulas for volume and surface of two– and three–dimensional symmetric objects are listed in the Math Review section *Symmetric Objects* on p. 82 in Chapter 3.

Solution to part (a):

$$\rho = \frac{1.0\ g}{1.0\ cm^3} = \frac{1.0\ g}{1.0 \times 10^{-6}\ m^3} = 1.0 \times 10^3\ \frac{kg}{m^3} \quad \textbf{(2)}$$

Thus, the mass of $1\ m^3$ of water is 1000 kg.

Solution to part (b):

$$m = \rho \cdot V = \rho \cdot \frac{4}{3}\pi \cdot r^3 = \frac{\rho}{6}\pi \cdot d^3 \quad \textbf{(3)}$$

with r the radius and d the diameter. Eq. [3] yields:

$$m_{bacterium} = 0.98 \cdot \frac{1000 \frac{kg}{m^3}}{6} \pi \cdot (1.0\ \mu m)^3 \qquad \textbf{(4)}$$

which results in $m_{bacterium} = 5 \times 10^{-16}$ kg.

Solution to part (c):

$$m = \rho \cdot V = \rho \cdot \pi \cdot r^2 \cdot l$$

$$= \left(1000 \frac{kg}{m^3}\right) \pi\ (1\ mm)^2\ (4\ mm) \qquad \textbf{(5)}$$

$$= 1.3 \times 10^{-5}\ kg = 0.013\ g$$

Problem 3.3

Fig. 3.51 shows a standard man intending to do *reverse curls* in a gym. The person holds the arms straight, using an overhand grip to hold the bar. If the mass of the bar is 100 kg, what is the tension in each of the shoulders? Consider the weight of the arm (see Table 3.3) and forces due to the weight of the bar.

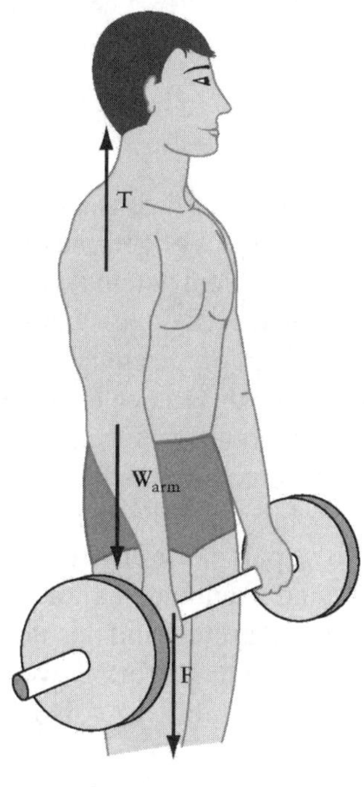

FIGURE 3.51

Solution: We treat the arm, not the entire person, as the object of interest. For a body part, forces acting across the interface to the rest of the body may have to be included as contact forces. In the particular case, focussing on the shown arm, three forces are included: a force due to the weight of the bar, **F**, the weight of the arm, $\mathbf{W}_{arm}$, and the tension force in the shoulder, **T**. For these three forces, Fig. 3.68 shows the free–body diagram. The two weight–related forces act downward and the tension is pulling the arm up.

We choose the vertical axis as positive in the upward direction. We also note that the problem text implies that the arm does not accelerate at the instant for which we solve the problem. Therefore, the problem is

Table 3.3

General data	
Age	30 years
Height	173 cm
Mass distribution	
Body mass M_{total}	70 kg
Mass of the trunk	48 %
Muscle mass	43 %
Mass of each leg	15 %
Fat mass	14 %
Bone mass	10 %
Mass of the head	7 %
Mass of each arm	6.5 %
Brain mass	2.1 %
Mass of both lungs	1.4 %
Homeostasis data	
Surface area	1.85 m²
Body core temperature*	310 K
Specific heat capacity	3.60 kJ/(kg K)
Respiratory data	
Total lung capacity	6.0 L
Tidal volume (lungs)	0.5 L
Breathing rate	15 breaths/min
Oxygen consumption	0.26 L/min
Carbon dioxide production	0.208 L/min
Cardiovascular data	
Blood volume	5.1 L
Cardiac output	5.0 L/min
Systolic blood pressure	16.0 kPa
Diastolic blood pressure	10.7 kPa
Heart rate	70 beats/min

* Skin surface temperatures vary with environmental temperature.

an application of Newton's first law. Using the free–body diagram in Fig. 3.68, we write for the vertical force components:

$$T - W_{arm} - F = 0 \qquad \textbf{(6)}$$

We use Eq. [6] to solve for the tension **T**. The two other forces are quantified with the information given in the problem text, Table 3.3 and Fig. 3.51. The magnitude of the force **F** is one–half of the weight of the bar. This is due to the fact that the person holds the bar with two hands evenly, i.e., each arm must support 50 % of its weight. The weight of the arm is calculated from Table 3.3:

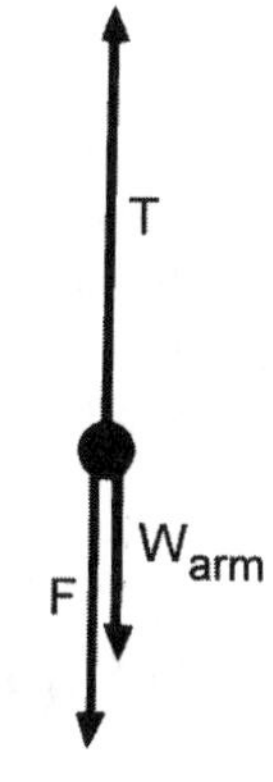

⇐ Figure 3.68

the mass of the arm is 7 % of 70 kg: M_{arm} = 4.9 kg. Thus, Eq. [6] reads:

$$T = W_{arm} + F$$
$$= \left(4.9\ kg + \frac{100\ kg}{2}\right) g = 538\ N \qquad \textbf{(7)}$$

Problem 3.5

Large hawks, eagles, vultures, storks, the white pelican and gulls are North American birds that sail on rising columns of warm air. This static soaring requires only 5 % of the effort of flapping flight. The birds are essentially in a level flight, holding their wings steadily stretched. The weight of the bird is balanced by a vertical lift force, which is a force exerted by the air on the bird's wings. How large is the lift force for

(a) a Franklin's gull (found in Alberta, Saskatchewan and Manitoba) with an average mass of 280 g, and

(b) an American white pelican (found in Western Canada) with an average mass of 7.0 kg?

⇐ Figure 3.69

Solution to part (a): The object of interest is the bird while sailing without flapping its wings. Four forces are involved in flight: thrust and drag in the horizontal direction, lift and weight in the vertical direction. To achieve flight at constant speed, the thrust must compensate the drag force. In the current problem, we are interested in levelled flight, which requires that the lift force compensates the weight. The free–body diagram in Fig. 3.69 shows the two forces acting in the vertical direction. Levelled flight implies that no vertical acceleration occurs, thus:

$$F_{lift} - W = 0 \qquad \textbf{(8)}$$

in which we have chosen the direction up as the positive axis. The weight is entered as a negative value as it is directed downward. Substituting the magnitude of the weight as $W = m \cdot g$ with m the given mass of a Franklin's gull, we find:

$$F_{lift} = m \cdot g = (0.28\ kg)\left(9.8\ \frac{m}{s^2}\right) = 2.74\ N \qquad \textbf{(9)}$$

Note that the mass of the bird has been used in standard unit kg. This is necessary because the standard unit of the resulting force, N, contains the unit kg not g.

Solution to part (b): The only difference in part (b) is the much larger mass of the American white pelican. Rewriting Eq. [9] with the mass of the pelican, we find F_{lift} = 68.6 N. Thus, a pelican needs a 25 fold higher lift force to sail than the small gull.

Problem 3.7

Fig. 3.54 shows two objects in contact on a frictionless surface. A horizontal force **F** is applied to the object with mass m_1.

(a) Use m_1 = 2.0 kg, m_2 = 1.0 kg and F = 3.0 N to calculate the magnitude of the force **f** between the two objects.

(b) Find the magnitude of the force **f** between the two objects if the force **F** is instead applied to the object of mass m_2 but in the opposite direction.

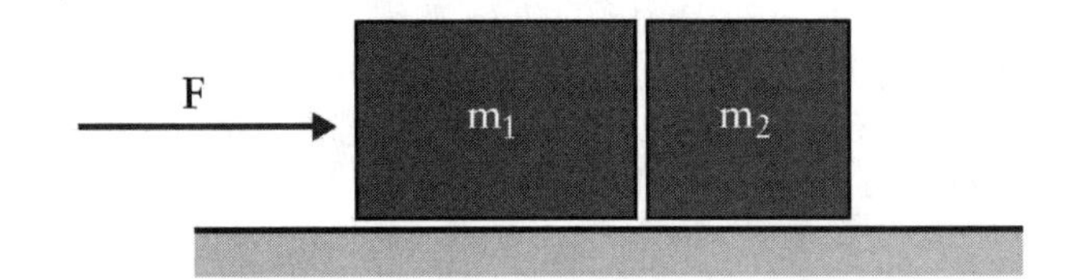

FIGURE 3.54

Solution to part (a): The frictionless surface is a part of the environment. Object 1 is the object of (primary) interest as all forces named in the problem act on it. Thus, we choose it as the system. This leaves open what role object 2 plays. It could either be part of the environment or it could be a second system. The difference is that we need to develop a free–body diagram for a system but not for a component of the environment. It is fair at this point to choose either role for object 2. However, if you choose to consider object 2 to be part of the environment you may have to come back and change that choice. If you choose object 2 to be a second system and this is not necessary, you invest some superfluous effort, however without adversely affecting your ability to solve the problem.

As it turns out, in the current case we have to identify object 2 as a second system, because we need some formulas derived from its free–body diagram to solve the problem. Consequently, several of the following steps have to be done twice, once for object 1 and once for object 2.

There are four forces acting on object 1, its weight, a normal force upward due to the contact with the frictionless surface, the external contact force **F** toward the right, and the interaction force (also a contact force) between the two objects, $\mathbf{f}_1$. This force is directed toward the left as it is exerted by object 2 on object 1. Only the two anti–parallel forces **F** and $\mathbf{f}_1$ are considered further because we are exclusively interested in effects along the horizontal surface.

Three forces act on object 2. These are its weight, the normal force due to the frictionless surface and the interaction force between the two objects, $\mathbf{f}_2$. Newton's third law relates the magnitudes of the two forces $\mathbf{f}_1$ and $\mathbf{f}_2$ to each other:

$$f_1 = f_2 \qquad \textbf{(10)}$$

These two forces point in opposite directions.

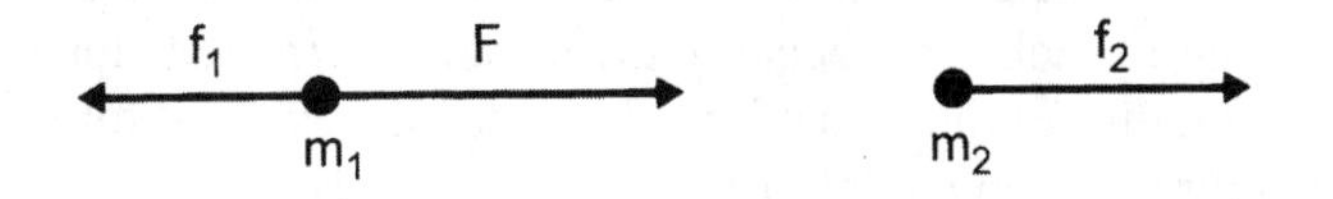

Figure 3.70

The free–body diagrams for objects 1 and 2 are shown in Fig. 3.70. We choose in both cases the positive axis toward the right. An acceleration is observed along this axis. Thus, Newton's second law applies. We use Fig. 3.70 to apply Newton's law to each system. For simplicity, we drop the indices of forces $\mathbf{f}_1$ and $\mathbf{f}_2$, re-writing Eq. [10] in the form $f_1 = f_2 = f$. This yields:

$$\begin{aligned} &\textit{system } 1: \quad F - f = m_1 \cdot a \\ &\textit{system } 2: \qquad\quad f = m_2 \cdot a \end{aligned} \qquad \textbf{(11)}$$

Note that the same acceleration a is assumed. This is justified because the two blocks always move together. Using the second formula in Eq. [11] to eliminate the acceleration a in the first formula, we find:

$$\begin{aligned} &a = \frac{f}{m_2} \quad \Rightarrow \quad F - f = \frac{m_1}{m_2} f \\ &f = \frac{F \cdot m_2}{m_1 + m_2} = \frac{(3.0\,N)\,(1.0\,kg)}{(1.0\,kg) + (2.0\,kg)} = 1.0\,N \end{aligned} \qquad \textbf{(12)}$$

Solution to part (b): In this part the same reasoning applies, except that the force **F** is this time applied to object 2 (i.e., objects 1 and 2 change places in the free–body diagrams of Fig. 3.70). When following this approach, we find instead of Eq. [12]:

$$f = F \frac{m_1}{m_1 + m_2} = 2.0\,N \qquad \textbf{(13)}$$

Why is the force in part (b) different in spite of the fact that the two bodies have the same acceleration a? The force **f** is responsible for accelerating the object on which the external force **F** does not act. In the second case the force **f** must accelerate the more massive object, and therefore, must be larger.

Problem 3.9

Fig. 3.56 shows two objects that are connected by a massless string. They are pulled along a frictionless surface by a horizontal external force. Using $F_{ext} = 50$ N, $m_1 = 10$ kg and $m_2 = 20$ kg, calculate

(a) the magnitude of the acceleration of the two objects, and

(b) the magnitude of the tension **T** in the string.

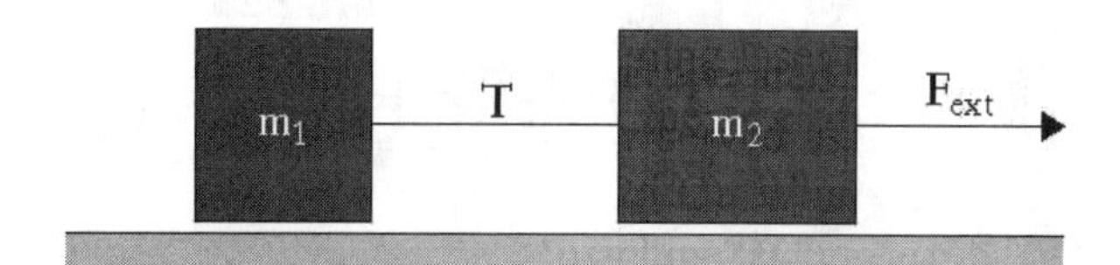

FIGURE 3.56

Solution: Even if there are several distinct objects present in a problem, it is not always necessary to treat them as separate systems. This problem is an example of this: we initially choose to treat the two connected objects together as a single system, and only in part (b) choose as system one of the objects separately.

Solution to part (a): Since the two objects are connected with a taut, massless string, they must move and accelerate together, which enables us to treat both objects as a single system. The external force is the only horizontal force acting on this system. We don't study forces in the vertical direction because the problem is confined to the horizontal direction. For a single force we do not need to draw a free–body diagram. We define the positive x–axis in the direction of the external force.

As an acceleration is explicitly mentioned, this problem is an application of Newton's second law. It reads for the x–component of the force:

$$F_{ext} = (m_1 + m_2)\, a \qquad \textbf{(14)}$$

which yields for the magnitude of the acceleration **a**:

$$a = \frac{50\ N}{(10\ kg) + (20\ kg)} = 1.67\ \frac{m}{s^2} \qquad \textbf{(15)}$$

Solution to part (b): We obtained in part (a) no information regarding the tension in the string between the objects. Therefore, the definition of the system must now be modified: we consider object 1 as the system and object 2 as part of its environment to solve for the tension **T**. This case leads again to a simple free–body diagram as the tension is the only horizontal force that acts on object 1. Keeping the same *x*–axis as in part (a), Newton's second law is written in the form:

$$T = m_1 \cdot a \qquad \textbf{(16)}$$

which yields:

$$T = (10\ kg)\left(1.67\ \frac{m}{s^2}\right) = 16.7\ N \qquad \textbf{(17)}$$

Problem 3.11

We study once more Example 3.4, except that the direction of the external force is changed to act tangentially to the circular path of the object of mass $m = 1.0$ kg, as shown in Fig. 3.58. We assume again that the object is held at a position where the string forms an angle $\theta = 30^0$ with the vertical.

(a) What is the magnitude of the tension **T** in the string?
(b) What is the magnitude of the force $\mathbf{F}_{ext}$?
(c) Why is it not possible to repeat the calculation with the external force $\mathbf{F}_{ext}$ acting vertically upward?

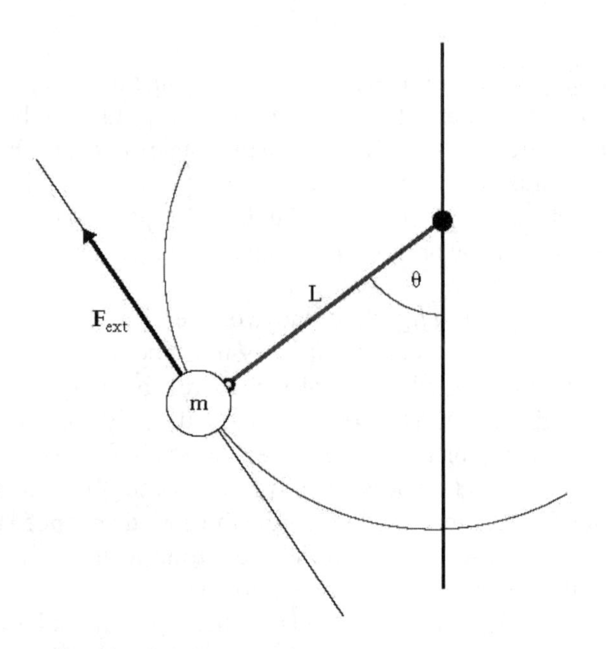

FIGURE 3.58

Solution to parts (a) and (b): The object of mass *m* attached to the string is the system; the string and the ceiling with the pivot point of the string are part of the environment. Three forces act on the object: its weight and two contact forces, the tension in the string and the external force.

Fig. 3.71 shows the free–body diagram for the object. The external force and the tension are perpendicular to each other because the tension is directed along the radius of the path of the object and the external force is directed tangential to that path. As a rule of thumb, we want to maximize the number of forces that act along the fundamental axes; thus, we choose the *y*–axis in the radial direction of the path and the *x*–axis in the tangential direction. With this choice, the tension in Fig. 3.71 has no *x*–component and its *y*–component is $T_y = +|\mathbf{T}|$, in which $|\mathbf{T}|$ indicates the magnitude of the vector **T**. In an analogous fashion, the external force in Fig. 3.71 has a zero *y*–component with the *x*–component $F_{ext,x} = -|\mathbf{F}_{ext}|$. This component is negative since the external force is directed opposite to the positive *x*–direction.

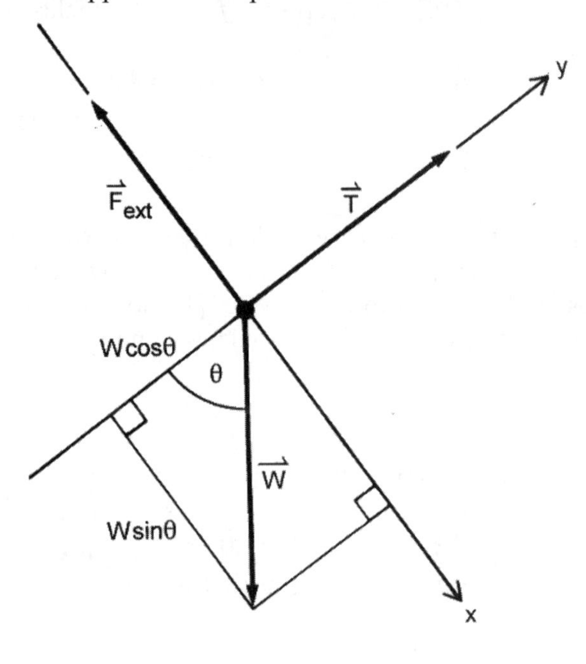

Figure 3.71

The weight **W** points vertically down and is therefore not parallel to either of the fundamental axes. We need to apply trigonometry to identify its two components: using the angle θ as given in Fig. 3.58 and shown again in Fig. 3.71, we find that $W_x = + W \cdot \sin\theta$ and $W_y = - W \cdot \cos\theta$.

This example illustrates that we often deal with a larger number of force components in a mechanical problem. It is important that you keep track of all these entries. One recommended approach is the preparation of Table 3.5, in which all force components are organized. Note that Table 3.5 also contains the sum of the respective force components, as required for the next step in the solution process. The object is held at rest by the external force. Thus, the problem is an application of Newton's first law. Using Fig. 3.71, we write:

$$\textit{x–direction:} \quad - F_{ext} + W \cdot \sin\theta = 0$$
$$\textit{y–direction:} \quad + T - W \cdot \cos\theta = 0 \qquad \textbf{(18)}$$

in which the first formula shows that the sum of the x–components of all forces is zero, and the second formula shows the same for the force components in the y–direction.

Solution to part (a): We use the second formula in Eq. [18] to solve for the magnitude of the tension:

$$T = W \cdot \cos\theta$$
$$= (1.0\ kg)\left(9.8\ \frac{m}{s^2}\right)\cos 30^0 = 8.5\ N \qquad \textbf{(19)}$$

Solution to part (b): We use the first formula in Eq. [18] to solve for the magnitude of the external force:

$$F_{ext} = W \cdot \sin\theta$$
$$= (1.0\ kg)\left(9.8\ \frac{m}{s^2}\right)\sin 30^0 = 4.9\ N \qquad \textbf{(20)}$$

Table 3.5

Force	x–component	y–component
Tension **T**	0	$+ \lvert\mathbf{T}\rvert = + T$
External force $\mathbf{F}_{ext}$	$- \lvert\mathbf{F}_{ext}\rvert = - F_{ext}$	0
Weight **W**	$+ W \cdot \sin\theta$	$- W \cdot \cos\theta$
Net force	$W \cdot \sin\theta - F_{ext}$	$T - W \cdot \cos\theta$

Solution to part (c): Consider the consequences a change in direction of the external force would cause in the free–body diagram of Fig. 3.71: the tension would become the only force with a non–vertical component. As a result, only one horizontal force component would act on the object; based on Newton's laws this cannot lead to a mechanical equilibrium. Thus, the object would have to accelerate, contrary to the implication in the problem text.

Problem 3.13

Fig. 3.60 shows the human leg (a) when it is stretched and (b) when it is bent. Note that the kneecap (3) is embedded in the quadriceps tendon (6) and is needed to protect the quadriceps tendon against wear and tear due to the femur (1) in the bent position. Assume that the magnitude of the tension in the quadriceps tendon of a bent knee is T = 1400 N. Use an angle $\theta_1 = 20^0$ between the horizontal and the direction of the tension above the kneecap and $\theta_2 = 10^0$ between the vertical and the direction of the tension below the kneecap, as shown in Fig. 3.61(a). What are the magnitude and the direction of the resultant force exerted on the femur, labelled – **R** in Fig. 3.61(b)?

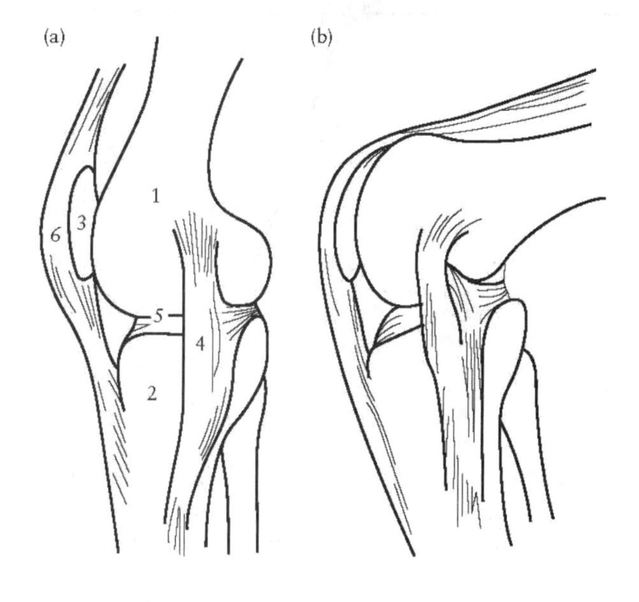

FIGURE 3.60

Solution: Even though we are asked to calculate a force acting on the femur, it is not possible to choose the femur as the system. To do so, we would need to include in the discussion several other major forces acting on the femur, such as forces acting at the hip joint. On the other side, we know most of the forces which act on the kneecap. If we choose the kneecap as the system, this leads us

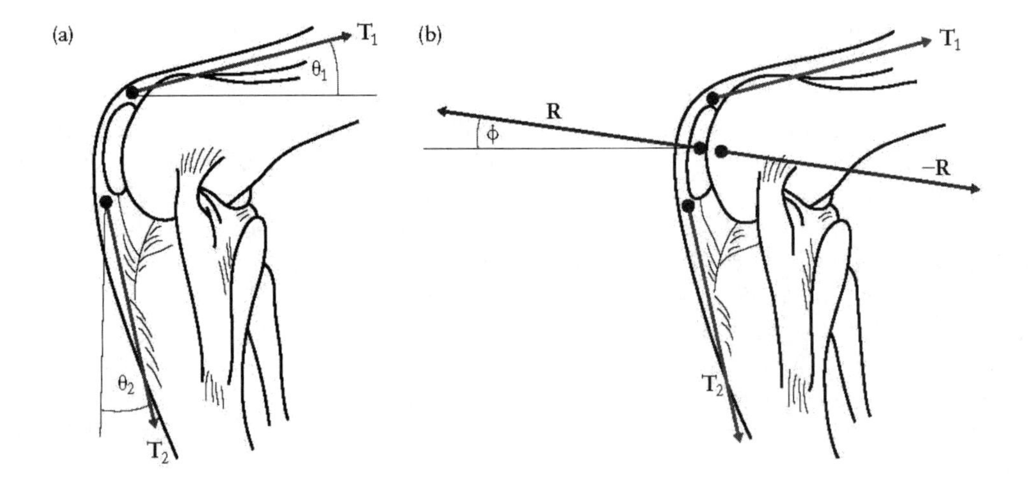

FIGURE 3.61

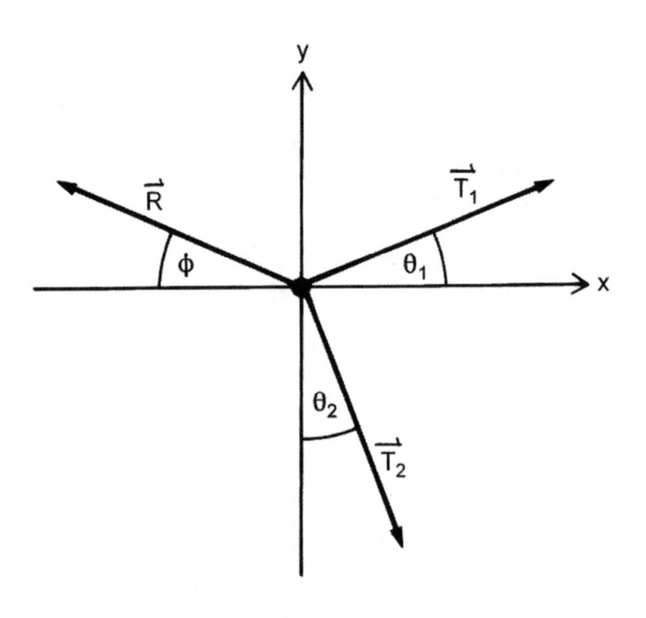

Figure 3.72

to an answer because one of the forces acting on the kneecap is due to the femur. Using Newton's third law, we will then be able to relate this force to the force we are asked to calculate.

Three forces act on the kneecap: two tension forces due to the quadriceps tendon, and the contact force exerted by the femur. Note that the two tensions are not the same as they have different directions. Thus, we follow the notation in Fig. 3.61(a) and label these two forces $\mathbf{T}_1$ and $\mathbf{T}_2$.

Fig. 3.72 shows the free–body diagram for the kneecap. The two angles θ_1 and θ_2 are the same as in Fig. 3.61(a). The free–body diagram further shows the force **R**, which is the contact force due to the femur. It is important to *not* draw this force in the wrong direction. We know that **R** is directed somehow toward the left to balance the components of the two tension forces which point toward the right. But we cannot assume that **R** is horizontal. Thus, the force **R** is chosen toward the left, forming a variable angle φ with the horizontal. If **R** turns out to be horizontal, the calculation will yield $\varphi = 0^0$. However, if $\varphi \neq 0^0$, drawing **R** horizontally in Fig. 3.72 would be a mistake which cannot be corrected in the later calculations.

Fig. 3.72 also shows our choice of coordinate system: the *x*–axis horizontally toward the right and the *y*–axis vertically upward. The problem is an application of Newton's first law in both directions because the kneecap does not accelerate either up or forward when the knee is bent. Using Fig. 3.72, we write Newton's laws for the kneecap:

x–direction:

$$T\cos\theta_1 + T\sin\theta_2 - R\cos\varphi = 0$$

y–direction:

$$T\sin\theta_1 - T\cos\theta_2 + R\sin\varphi = 0 \qquad \textbf{(21)}$$

Note that we wrote T in Eq. [21] for the magnitude of the tension as T_1 and T_2 are the same. Eq. [21] contains two unknown parameters: R and φ. Together they provides us with **R**. To solve Eq. [21], we isolate the unknown parameter φ on the right hand side of each formula:

x–direction:

$$T\cos\theta_1 + T\sin\theta_2 = R\cos\varphi$$

y–direction:

$$-T\sin\theta_1 + T\cos\theta_2 = R\sin\varphi \qquad \textbf{(22)}$$

Now we divide the two formulas (the second divided by the first formula):

$$\tan\varphi = \frac{-\sin\theta_1 + \cos\theta_2}{\cos\theta_1 + \sin\theta_2} \quad \textbf{(23)}$$

Next we substitute the given angles in Eq. [23]:

$$\tan\varphi = \frac{-\sin 20^0 + \cos 10^0}{\cos 20^0 + \sin 10^0} = 0.577 \quad \textbf{(24)}$$

which yields $\varphi = 30^0$. The magnitude of the force R is obtained by substituting the result from Eq. [24] in either one of the two formulas in Eq. [22]. Choosing the first formula, we find:

$$R = \frac{T(\cos\theta_1 + \sin\theta_2)}{\cos\varphi} = \frac{(1400\ N)(\cos 20^0 + \sin 10^0)}{\cos 30^0} = 1800\ N \quad \textbf{(25)}$$

Two closing comments: first, note that **R** is a very large force. You need a force of this magnitude to lift an object of a mass of about 185 kg. While bending our knees is done routinely, one can easily imagine how such forces lead to injury or wear and tear.

Secondly, note that Eqs. [24] and [25] are not the answer to the problem. In a last step, Newton's third law must be applied. This is illustrated in Fig. 3.61(b): we determined the force **R** acting on the kneecap, but we sought the force **–R** acting on the femur. Thus, the answer is that a force of magnitude of 1800 N acts on the femur at an angle 30^0 below the horizontal.

Figure 3.63

Problem 3.15

In Fig. 3.63, a standard man does *cable crossover flys* in the gym. In this exercise, the person holds two handles with the arms spread, the chest slightly leaning forward and the elbow slightly bent. During the action, the handles are pulled forward until the hands touch; this contraction is accompanied by a lifting of two objects of equal masses M = 20 kg that are connected to the handles via pulleys (not shown). If the standard man locks the hands together at maximum contraction, what angle θ must the cable form with the vertical for the standard man to apply a force of 300 N vertically downward? Neglect the mass of the handles.

Solution: We choose the two locked handles as the system. Three forces act on this system: the two tension forces along the cables, $\mathbf{T}_1$ and $\mathbf{T}_2$, and the external force that the person applies downward on the handles, $\mathbf{F}_{ext}$. We do not include the weight of the handles because their mass is negligible.

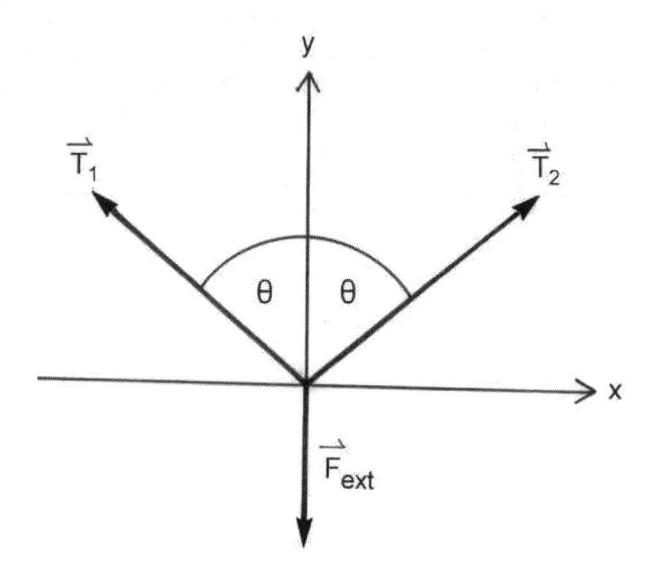

Figure 3.73

Fig. 3.73 shows the free–body diagram for the combined handles. We use Newton's first law for the net force component in the vertical direction because the handles do not accelerate in this problem:

$$-F_{ext} + T_1\cos\theta + T_2\cos\theta = 0 \quad \textbf{(26)}$$

The two tensions are each equal in magnitude to the weight of the masses M hanging over the pulleys: $W_M = T_1 = T_2$. Thus, Eq. [26] simplifies to:

$$-F_{ext} + 2 \cdot M \cdot g \cdot \cos\theta = 0 \quad \textbf{(27)}$$

Eq. [27] is solved for the angle θ:

$$\cos\theta = \frac{F_{ext}}{2 \cdot M \cdot g} = \frac{300\ N}{2\ (20\ kg)\left(9.8\frac{m}{s^2}\right)} = 0.765 \quad \textbf{(28)}$$

which yields $\theta = 40^0$.

Problem 3.17

Fig. 3.65 shows the top view of an object on a frictionless surface. Two horizontal forces act on the object. The force $\mathbf{F}_1$ acts in the positive x–direction and has a magnitude of 10 N; the force $\mathbf{F}_2$ acts toward the lower left (third quadrant). If the object accelerates in a direction which forms an angle $\theta = 30^0$ with the negative y–axis and has magnitude $a = 10\ m/s^2$, calculate the force $\mathbf{F}_2$
(a) in component notation, and
(b) as a magnitude and direction.

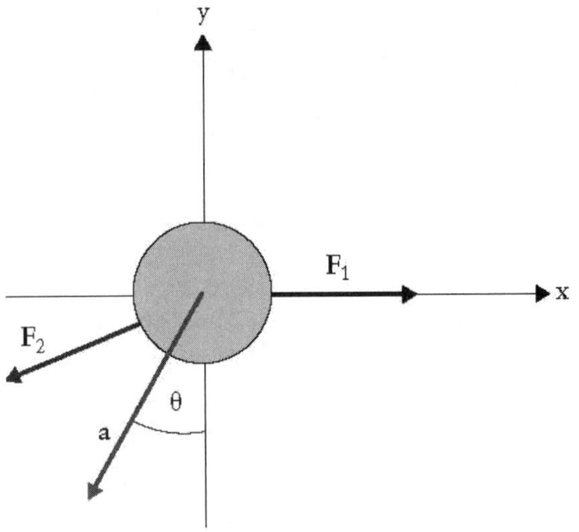

FIGURE 3.65

Solution to part (a): The circular box in Fig. 3.65 is the system. None of the components in the environment are shown, although they must be present because they are the origins of the forces referred to in the problem. The text identifies two forces acting on the object, $\mathbf{F}_1$ and $\mathbf{F}_2$. Of these, $\mathbf{F}_2$ is unknown (even though it is sketched in Fig. 3.65). We do not include the weight in this problem because it acts in the direction perpendicular to the plane shown in the top–view in Fig. 3.65.

The free–body diagram for this problem, which is not shown, contains only forces $\mathbf{F}_1$ and $\mathbf{F}_2$. Note that Fig. 3.65 is not a substitute for the free–body diagram: never add an acceleration to a free–body diagram as it isn't one of the forces but represents the term on the other side of Newton's second law, $\mathbf{F}_{net} = m \cdot \mathbf{a}$! The coordinate system has already been defined in Fig. 3.65.

The system is not in mechanical equilibrium since an acceleration has been identified. Thus, we use Newton's second law in component form:

$$\begin{aligned} &x\text{–direction:} \quad F_{1,x} + F_{2,x} = m \cdot a_x \\ &y\text{–direction:} \quad F_{1,y} + F_{2,y} = m \cdot a_y \end{aligned} \quad \textbf{(29)}$$

The two formulas in Eq. [29] are rewritten to determine the two components of the unknown force $\mathbf{F}_2$:

$$\begin{aligned} F_{2,x} &= m \cdot a_x - F_{1,x} \\ F_{2,y} &= m \cdot a_y - F_{1,y} \end{aligned} \quad \textbf{(30)}$$

in which $\mathbf{F}_1 = (10.0\ N, 0)$ is given in the problem text, and

$$\begin{aligned} m \cdot a_x &= -(1.5\ kg)\left(10\ \frac{m}{s^2}\right)\sin 30^0 \\ m \cdot a_y &= -(1.5\ kg)\left(10\ \frac{m}{s^2}\right)\cos 30^0 \end{aligned} \quad \textbf{(31)}$$

Both sides of Eq. [31] carry the unit $kg \cdot m/s^2$ which is the same as the unit N. Substituting the components of the force $\mathbf{F}_1$, and Eq. [31] in Eq. [30], we find:

$$\begin{aligned} F_{2,x} &= -17.5\ N \\ F_{2,y} &= -13.0\ N \end{aligned} \quad \textbf{(32)}$$

Solution to part (b): Fig. 3.74 illustrates how the two components of force $\mathbf{F}_2$ in Eq. [32] are related to its magnitude and direction. The magnitude is obtained with the Pythagorean theorem from Eq. [32]:

$$F_2 = \sqrt{(-17.5\ N)^2 + (-13.0\ N)^2} = 21.8\ N \quad \textbf{(33)}$$

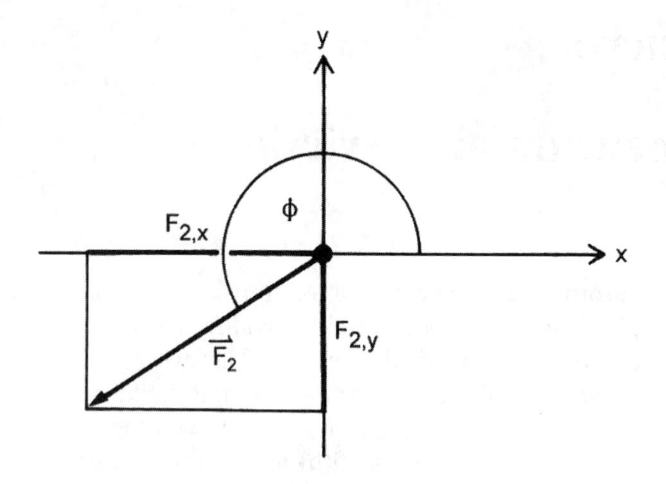

Figure 3.74

The direction of a vector is expressed by its angle with the positive x–axis, measured in the counter–clockwise direction. This angle is shown in Fig. 3.74 as angle φ. To obtain φ, we first calculate the angle ψ between the vector $\mathbf{F}_2$ and the negative x–axis. Using Eq. [32] and the trigonometric definition of the angle ψ, we get:

$$\tan\psi = \frac{F_{2,\,y}}{F_{2,\,x}} = \frac{-13.0\ N}{-17.5\ N} = +0.743 \qquad \textbf{(34)}$$

This corresponds to either $\varphi = 36.6^0$ or $\varphi = 216.6^0$, because there are two values for the trigonometric function. We choose $\varphi = 216.6^0$ because the vector lies in the third quadrant.

Problem 3.19

An object of mass $m = 6.0$ kg accelerates with 2.0 m/s². (a) What is the magnitude of the net force acting on it? (b) If the same force is applied on an object with mass $M = 4.0$ kg, what is the magnitude of that object's acceleration?

Solution to part (a): $F_{net} = m \cdot a = 12$ N.

Solution to part (b): $a = F_{net}/M = 3$ m/s².

Problem 3.21

The leg and cast in Fig. 3.67 have a mass of 22.5 kg. Determine the mass of object 2 and the angle θ needed in order that there be no force acting on the hip joint by leg plus cast. Note that object 1 has a mass of 11 kg and $\varphi = 40^0$.

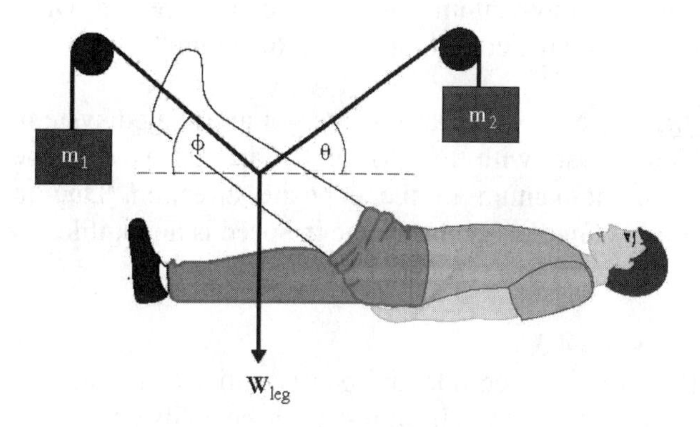

FIGURE 3.67

Solution: The mechanical equilibrium in the horizontal direction yields:

$$m_2 \cdot g \cdot \cos\theta - m_1 \cdot g \cdot \cos\varphi = 0 \qquad \textbf{(35)}$$

and in the vertical direction:

$$m_2 \cdot g \cdot \sin\theta + m_1 \cdot g \cdot \sin\varphi - W_{leg} = 0 \qquad \textbf{(36)}$$

This yields: $\theta = 61.4^0$ and $W_2 = 172.5$ N.

CHAPTER FOUR

Locomotion II: Momentum and Friction

MULTIPLE CHOICE AND CONCEPTUAL QUESTIONS

Question 4.1

Stand perfectly still, then take a step forward. Before the step your momentum was zero, then it increased. Does this violate the conservation of momentum?

Answer: No. Note that you are not an isolated system. You interact with the surface of Earth; it takes up the same momentum in the opposite direction. Due to Earth's huge mass, its change in speed is negligible.

Question 4.3

If two objects collide and one is initially at rest, is it possible for both to be at rest after the collision?

Answer: No. Their combined final momentum is zero, but for a collision to take place the second object must move with an initial non–zero speed.

Question 4.5

Fig. 4.19 shows two escape options for a squirrel (solid circle) that is chased by a dog (small open circle) in a garden: run straight for the closest tree (large open circle) and up (Fig. 4.19(a)), or run along a tangent toward the next tree and then up on the far side (Fig. 4.19(b)). Why does the squirrel usually chose the second option? *Hint*: consider the collision squirrel–tree and a possible collision dog–squirrel when the squirrel is runing up the tree.

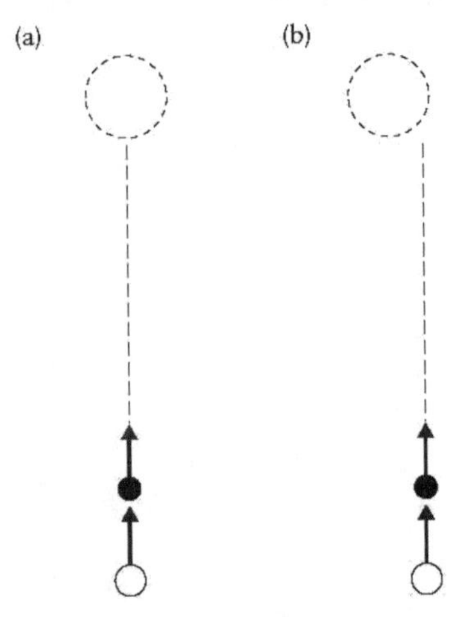

FIGURE 4.19

Answer: In both collisions the squirrel is much lighter than the other object and would have to accommodate a major change in momentum. When reaching the tree in Fig. 4.19(a) the squirrel's leg muscles and tendons would have to absorb a negative acceleration to rest in the horizontal direction, while concurrently providing a positive acceleration in the vertical direction. This would slow the progress of the escape while the dog could continue along a straight line toward collision with the squirrel. The approach in Fig. 4.19(b) allows for a longer path segment during which the squirrel adjusts its horizontal and vertical velocity components.

With the dog heavier than the squirrel, the dog's preferred strategy is a direct collision with the squirrel. Were the squirrel to run up the tree such that the dog could launch toward it, the dog's chance to achieve such a collision is significantly enhanced over the squirrel choosing a path that eliminates a collision once the squirrel reaches the foot of the tree.

Question 4.7

Velvet worms use a rather cruel method of hunting: aiming at their prey for distances of up to one metre, they eject two jets of a sticky liquid from glands near their mouth. The liquid dries very fast, immobilizing the prey. The worm then injects digestive juices through a bite and later extracts its meal, which formed by internal liquefaction of the prey. Would not the conservation of linear momentum require the velvet worm's body to be accelerated backward while spraying from its two powerful nozzles?

Answer: No. The velvet worm is not an isolated system, but holds steady to the ground. Thus, the backward acceleration affects the worm and Earth it holds on to, which leads to a negligible acceleration. This situation is similar to the recoil on a person firing a gun while kneeling on steady ground.

Question 4.9

A 1.2-kg common raven, just leaving a feeding platform at 0.2 m/s, and a 23-g Savannah sparrow in mid–flight at 10 m/s, collide with a glass window (treat the collision as a perfectly inelastic collision). Which bird exerts the larger force on the window panel? (A) the raven; (B) the sparrow; (C) they exert the same force; (D) neither exerts a force; (E) we cannot answer the question with the information given.

Answer: (A). The raven's linear momentum is p_{raven} = 0.24 kg · m/s and the sparrow's is $p_{sparrow}$ = 0.23 kg · m/s. Let's assume that the window withstands the impact and that the collisions occur over the same amount of time. In an inelastic collision, linear momentum is reduced to zero during the collision if the window panel withstands the impact. That means that the raven exerts the larger force, though only slightly larger than the sparrow. If the collision is not inelastic, the momentum change is even greater as the birds bounce back. The respective forces are proportional to the linear momentum changes.

Question 4.11
Smaller birds, such as the European starling (introduced in 1890 in North America and now common across the continent) are often seen resting on small tree branches. When they take off, the branch swings momentarily downward. Can this observation be explained by the momentum disc theory?

Answer: No. The momentum disc theory deals only with the interaction of thrust and lift forces and the flow of air through the stroke plane. This model is not suitable to describe the take–off phase because the animal is in contact with a solid surface. Indeed, birds and most insects jump off the supporting surface to obtain an initial upward momentum to facilitate a successful take–off. This is important since the first complete cycle of flapping does not provide the same lift as subsequent cycles. Bats avoid the take–off stage altogether by hanging upside down while resting. In this case, gravity provides the necessary initial acceleration.

Question 4.13
A baseball player accidentally releases the bat in mid–swing. The bat sails toward the stands, spinning in the air. Neglect air resistance.
(a) With the information given, can we determine any point on the baseball bat for which we can describe its motion after it left the player's hands? What motion does that point perform?
(b) Assuming that the baseball bat is made of solid wood (uniform density), where along the baseball bat would you expect the point in part (a) to lie?

Answer to part (a): The centre of mass of the baseball bat follows Eq. [4.49]. The only external force on the baseball bat is gravity; we neglect air resistance and the player no longer has contact with the bat. Thus, the centre of mass moves along a parabola (projectile motion).

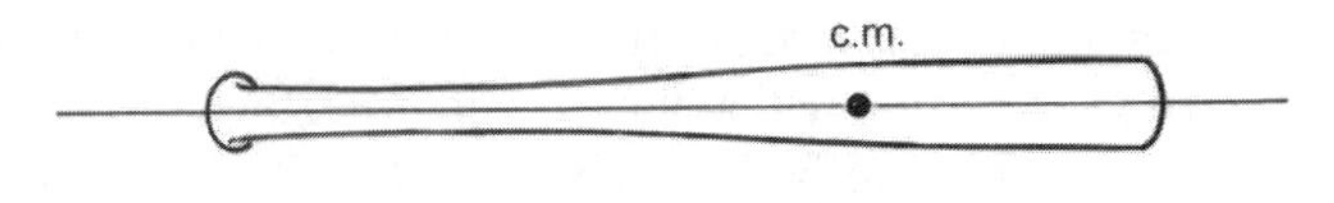

Figure 4.26

Answer to part (b): Fig. 4.26 shows a sketch of a baseball bat. The centre of mass lies closer to the massive end of the bat. If you own a baseball bat, confirm this by measuring its length and then support the bat at the half–way point. The more massive end tilts down and the bat falls to the ground. To balance it properly, you need to support it at the centre of mass position that is closer to the massive end. Convince yourself that the mass of the bat left and right of the centre of mass position is not the same (which is a confusing fact because the label "centre of mass" appears misleading).

Question 4.15
A system consists of three objects. Initially its centre of mass moves with constant speed along a straight line. What happens if I exert an external force on only one of the three objects? (A) Nothing; to change the state of motion I must exert the external force on all three objects; (B) Nothing; I cannot exert a force on just a part of a system; (C) Nothing; I need collisions among the three objects to affect the centre of mass; (D) The object I interact with accelerates, but the other two objects accelerate in the opposite direction such that the centre of mass continues to move at constant speed; (E) All three objects accelerate to allow the centre of mass to accelerate in the same fashion as the object on which I exert the external force. This is necessary because otherwise the external force causes the dismantling of the system, which is inconsistent with the definition of the system as three objects; (F) None of the above. If you choose (F), would you have a better suggestion?

Answer: (F). We exert an unbalanced external force on the system, thus its centre of mass must accelerate. This acceleration can be the result of just one object's motion relative to the others. Inspect Eqs. [4.47] and [4.49] to rule out the first five choices given.

Question 4.17
Why do car manufacturers offer ABS systems with their products?

Answer: Static coefficients of friction are always larger than the kinetic coefficients of friction at the same interface. The interaction of a locked (i.e., sliding) tire with the road surface is described by kinetic friction while the interaction of a rolling tire is closer to a static friction case.

Question 4.19
We compare the terminal speed of a raindrop of mass 30 mg and an equally heavy ice pellet during a hail storm. Noting the density difference, with a value of $\rho = 1.0$ g/cm³ for liquid water and $\rho = 0.92$ g/cm³ for ice,
(a) do both fall with the same terminal speed (since the density of the falling object is *not* a factor in Eq. [4.52]), and
(b) if they fall with different terminal speeds, which one is faster?

Answer to part (a): The terminal velocity of an object falling through air is given in Eq. [4.52]:

$$v_{terminal} = \sqrt{\frac{2 \cdot m \cdot g}{D \cdot \rho \cdot A}} \qquad (1)$$

The density of the medium, ρ, does not vary if the temperature of the medium is fixed. Note that this density refers to the density of air in this example, not the density of the raindrop or ice pellet. We find the same drag coefficient D because rain drops and hail pellets are of spherical shape. Thus, Eq. [1] is rewritten as the ratio of the two terminal speeds in the form:

$$\frac{v_{drop,\ terminal}}{v_{hail,\ terminal}} = \frac{\sqrt{\dfrac{m_{drop}}{A_{drop}}}}{\sqrt{\dfrac{m_{hail}}{A_{hail}}}} = = \sqrt{\frac{m_{drop} \cdot A_{hail}}{A_{drop} \cdot m_{hail}}} \qquad (2)$$

The problem states that both objects have the same mass. Using the definition of density, $\rho = m/V$, and the volume of a sphere $V = 4 \cdot \pi \cdot r^3/3$, we find the relation of the radii of raindrop and hail pellet:

$$\rho_{drop} \cdot V_{drop} = \frac{4}{3} \cdot \pi \cdot r_{drop}^3 \cdot \rho_{drop}$$
$$= \rho_{hail} \cdot V_{hail} = \frac{4}{3} \cdot \pi \cdot r_{hail}^3 \cdot \rho_{hail} \qquad (3)$$

which yields:

$$r_{drop} = \left(\frac{\rho_{hail}}{\rho_{drop}}\right)^{1/3} r_{hail} \qquad (4)$$

We substitute this result in Eq. [2], using for the cross–sectional area the formula $A = \pi \cdot r^2$:

$$\frac{v_{drop,\ terminal}}{v_{hail,\ terminal}} = \sqrt{\frac{\pi \cdot r_{hail}^2}{\pi \cdot r_{drop}^2}}$$
$$= \frac{r_{hail}}{r_{drop}} = \left(\frac{\rho_{drop}}{\rho_{hail}}\right)^{1/3} \qquad (5)$$

with the given ratio of densities, we find:

$$\frac{v_{drop,\ terminal}}{v_{hail,\ terminal}} = \left(\frac{1.0}{0.92}\right)^{1/3} = 1.03 \qquad (6)$$

Thus, the two velocities are not the same because both objects of same mass differ in their volume and, therefore, in their cross–sectional area during motion through the air.

Answer to part (b): Eq. [6] states that the raindrop falls faster, even though there is only a 3 % difference in terminal speed.

Question 4.21
Why is the coefficient of kinetic friction smaller than the coefficient of static friction?
(a) Give a reason based on the microscopic roughness of the interface, and
(b) give a logical reason by looking at Eqs. [4.54] and [4.56] and thinking about what they imply.

Answer to part (a): Friction is caused at the microscopic level by electric interactions between atoms of the adjacent surfaces, and by the roughness of both surfaces causing interlocking of protrusions. When the two surfaces are at rest relative to each other (static friction case) both the electric and morphological surface features settle into the most stable state (which we will define as a state of minimum potential energy in Chapter 6 for mechanical systems and in Chapter 13 for electric systems). Setting the system into motion out of this state means that we separate both surfaces while in a stable state. This requires a larger force than sliding them past each other, when only a small fraction of stronger interactions form across the surface at any given time.

Answer to part (b): In classical philosophy and mathematics, but rarely in scientific inquiry, statements can be verified by assuming that they are not true, then proving this assumption to be inconsistent with other observations. To demonstrate this approach, let's assume that there is a system for which $\mu_s < \mu_k$. We use Eqs. [4.54] and [4.56] to describe the resulting motion as we tilt the underlying surface to steeper and stepper angles; we specifically follow the same arguments we made in Example 4.6: up to angle θ_1 the system is governed by Eq. [4.54], correctly predicting that the two objects remain at rest relative to each other. When we now tilt the surface beyond angle θ_1, Eq. [4.54] requires the two objects to move relative to each other because the threshold of the static friction force is exceeded. More specifically, the dashed force in Fig. 4.27(a) no longer exceeds the force component W_x that points downhill, leading to an acceleration. At that moment we switch to Fig. 4.27(b) with the kinetic friction force $\mathscr{F}_k$. With our assumption above, this force will exceed the force component W_x for a range of tilt angles θ_1 to θ_2 with $\theta_2 > \theta_1$. In that interval, the motion is decelerated, i.e., the object comes to a rest. At that point, Eq. [4.54] applies again, requiring the object to accelerate. Thus, with the assumption $\mu_s < \mu_k$ we can show that Eqs. [4.54] and [4.56] become mutually contradictory in a range of angles for Fig. 4.27. The only way to resolve this problem is to require $\mu_s \geq \mu_k$.

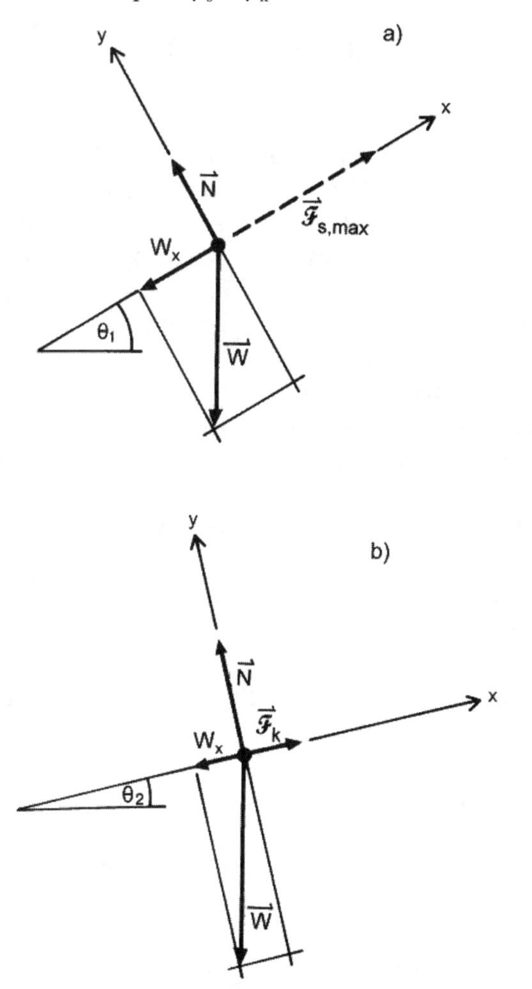

Figure 4.27

We compare the answers in part (a) and (b) to see why scientists don't like to use the mathematical ap-proach of arguing: in part (a) we learn new details about friction, in part (b) we only confirm that a given fact is the case. Gathering evidence for a scientific statement allows research to proceed and new knowledge to be uncovered; in the current case we develop a greater sense of relevance of the microscopic nature of interfaces. The answer in part (b) does not allow us to learn more about the underlying principles of the phenomenon.

Question 4.23

We discussed the case of objects moving through a liquid at small speeds in which the drag force is proportional to the speed v (linear relation). With the drag force being the only force acting in the horizontal direction on a fish that stops propelling its body forward, we expect that (A) The fish slows down linearly, $v \propto -t$ (t is time); (B) The fish slows down until it reaches a finite terminal speed; (C) The fish slows down but not with a linear relation between its speed and time; (D) The fish comes instantaneously to rest; (E) The fish slows down and is then pushed in the opposite direction by the drag force.

<u>*Answer*:</u> (C). This example shows that the term "linear" has to be used with some caution. In particular, if Newton's second law leads to a linear equation it usually will not have a linear function as the solution $v(t)$. In the current case the horizontal equation of motion for the fish is:

$$F_{net} = -k \cdot v = m \cdot a \qquad \textbf{(7)}$$

with the weight and the buoyant force the only other forces acting on the fish, but both in the vertical direction. Eq. [7] is called a *differential equation* because the acceleration is the change of the speed with time, $a = \lim \Delta v / \Delta t$ with the limit for $\Delta t \rightarrow 0$. In differential calculus, Eq. [7] is then written as:

$$-k \cdot v = m \frac{dv}{dt} \qquad \textbf{(8)}$$

We don't discuss the solution to this equation here, but if you are interested and taking an introductory level calculus course, you should be able to use the method of separation of the variables with a subsequent integration to find the solution of Eq. [8]. We comment on Eq. [8] a bit further in Problem 4.14.

A complete solution of Eq. [8] is not necessary to decide whether the solution is linear. For this, we just substitute a linear solution and see whether it is consistent with this equation. Such a linear function has the general form:

$$v(t) = a + b \cdot t \quad ; \quad a \text{ and } b \text{ constant} \qquad \textbf{(9)}$$

We further note that $\Delta v/\Delta t$ for Eq. [9] leads to a constant value (slope of the linear function); i.e., $\Delta v/\Delta t = b$. Substituting this observation and Eq. [9] in Eq. [7] yields:

$$-k(a + b \cdot t) = m \cdot b \qquad \textbf{(10)}$$

We see now that Eq. [9] is not a solution to Eq. [7] because Eq. [10] is incorrect: the left hand side of the equation varies with time but the right hand side is time–independent.

Choices (B) and (E) are inconsistent with Eq. [9] because the drag force becomes only zero when the fish is at rest. Choice (D) requires an infinite drag force, that could not be overcome by the fish in the first place.

Question 4.25

(a) When we refer to the shape of an animal as being aerodynamic, what formula — and particularly what *term* in that formula — do we have in mind?

(b) What is the underlying physics of an aerodynamic shape?

Answer to part (a): We refer to Eq. [4.50] that describes the motion of an object through a medium at high speed:

$$F_{drag} = \frac{D}{2}\rho \cdot A \cdot v_{system}^2 \qquad \textbf{(11)}$$

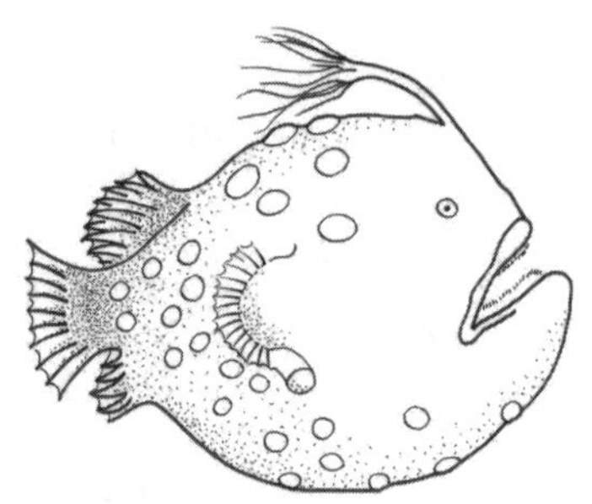

Figure 4.28

Figure 4.29

Note that aerodynamics does not relate to A which is the cross–section of the object perpendicular to the direction of its velocity vector: an *Atlantic Football Fish* (Fig. 4.28), which is a deep–sea angler fish of 60 cm length and a *Blue Marlin* (Fig. 4.29), one of the fastest of all fish with a long beaklike nose with a length of about 4 metres, can both have the same cross–sectional area A. However, the drag coefficient D quantifies the streamlining of the animal which determines its aerodynamic properties.

Answer to part (b): With aerodynamics we refer physically to the ratio v_{medium}/v_{system}, i.e., the speed of the moving animal relative to the speed to which the displaced medium is accelerated. The football–shaped angler fish must accelerate the water in front of its body to about the same speed sideways as its forward speed, otherwise it couldn't move through the dense medium. The Blue Marlin's streamlined body shape allows the sideways motion of the displaced water to be much slower than the fish's forward speed.

ANALYTICAL PROBLEMS

Problem 4.1

Polonium–210 (^{210}Po) is a radioactive isotope that has an atomic nucleus with 84 protons and 126 neutrons (together 210 nucleons). It undergoes an α–decay with a half–life of 138.4 days, resulting in a stable Lead–206 (^{206}Pb) isotope of a mass of 205.974 · u. Assuming that the polonium nucleus is at rest when it decays, an α–particle (^4He) with a speed of 1.6×10^7 m/s is emitted. The α–particle has a mass of 4,002 · u. The atomic unit u is defined as $1\ u = 1.6605677 \times 10^{-27}$ kg.

(a) In what direction does the daughter isotope ^{206}Pb move?

(b) What is the speed of the ^{206}Pb leaving the decay zone?

(c) Compare both speeds in this problem to the vacuum speed of light.

Solution to part (a): Conceptually, the conservation of momentum requires that the lead nucleus and the α–particle move in opposite directions. We define this common line of motion as the x–axis.

Solution to part (b): We use the x–components in Eq. [4.5] with const = 0 for the initial state (particle at rest):

$$m_{Pb} \cdot v_{Pb,\,final} + m_\alpha \cdot v_\alpha = 0 \qquad \textbf{(12)}$$

This is solved for the final speed of the Pb nucleus:

$$v_{Pb,\,final} = -\frac{m_\alpha}{m_{Pb}} v_\alpha = -\frac{4.002\ u}{205.974\ u}\left(1.6 \times 10^7\ \frac{m}{s}\right) = -3.1 \times 10^5\ \frac{m}{s} \qquad \textbf{(13)}$$

The lead nucleus moves with 3.1×10^5 m/s in the direction opposite to the α–particle.

<u>Solution to part (c):</u> Both speeds are significantly below the speed of light in vacuum with the α–particle moving at 5 % of the speed of light and the lead particle at 1 ‰. Note that we would have to be more careful if any of the calculated speeds came closer to the speed of light; classical mechanics fails for objects moving with a significant fraction of the speed of light and has to be replaced by relativistic calculations.

Problems 4.3

Some birds of prey, such as the Northern Goshawk, hunt other birds in midair. Typically, the hawk spots the prey while soaring high above, then dives in for the kill with the deadly grip of its talons. They do not shy away from rather large prey, such as ducks and crows. We want to estimate the impact on the hawk when completing a successful kill by using the change in linear momentum when clawing into the prey, which we take to be an unsuspecting American crow. Use: mass of crow $m_C = 450$ g, horizontal flight of the crow with speed $v_C = 10$ m/s, mass of hawk $m_H = 900$ g, flight path in the same plane as crow but approaching from behind with $v_H = 20$ m/s and with an angle of $\theta = 70^0$ with the horizontal, as illustrated in Fig. 4.21.

(a) How fast does the hawk move immediately after catching the crow?

(b) By what angle has the direction of motion of the hawk changed at the impact? *Note*: American crows take the danger posed by high–soaring hawks very seriously. They are often seen to mob the predator as a group to drive it out of their neighbourhood.

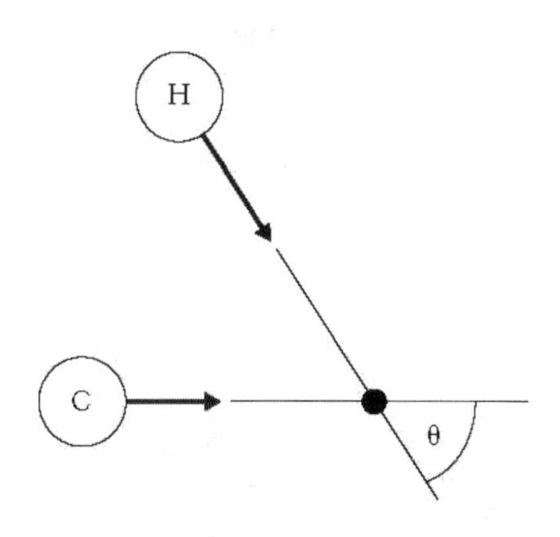

FIGURE 4.21

<u>Solution to part (a)</u>: This problem is similar to the previous one, except that we need to consider two dimensions which describe the plane in which the hawk and the crow move, as shown in Fig. 4.21. We call this plane the *xy*–plane and use the first two formulas in Eq. [4.11] with index "i" for initial and "f" for final:

$$\begin{aligned} &x\text{–direction:} \\ &\quad m_H v_{H,i,x} + m_C v_{C,i,x} = (m_H + m_C) v_{f,x} \\ &y\text{–direction:} \\ &\quad m_H v_{H,i,y} = (m_H + m_C) v_{f,y} \end{aligned} \qquad \textbf{(14)}$$

in which we used $v_{C,i,y} = 0$ because the crow flies horizontally.

A note of caution at this point of the calculations. With the large number of parameters in these collision problems, there is a temptation to replace variables with their respective numerical values in Eq. [14]. This is not advisable though because it leads to confusing mixtures of variables, numbers and units. Instead, perform all calculations with the variables as given in Eq. [14] and substitute numerical values only when no further mathematical operations are applied.

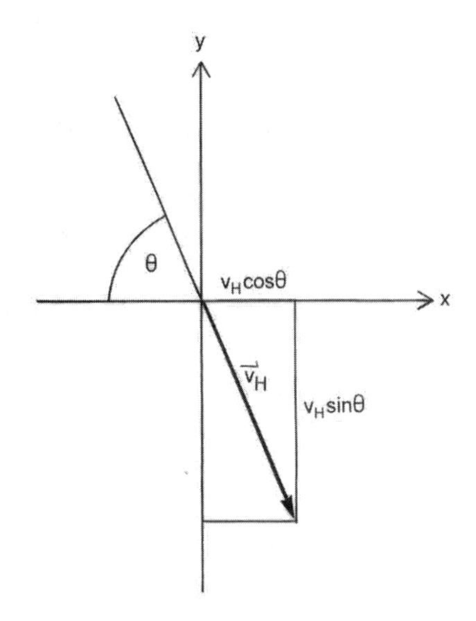

Figure 4.30

We choose the positive x–axis toward the right and the positive y–axis upward. Fig. 4.30 shows the x– and y–components of the velocity vector of the hawk. From Fig. 4.30 and the problem text we find:

$$v_{C,i,x} = v_C$$
$$v_{H,i,x} = v_H \cos\theta \quad ; \quad v_{H,i,y} = -v_H \sin\theta \qquad \textbf{(15)}$$

we substitute the three terms in Eq. [25] into the two formulas in Eq. [24]:

x–direction:
$$m_H v_H \cos\theta + m_C v_C = (m_H + m_C) v_{f,x}$$
y–direction:
$$-m_H v_H \sin\theta = (m_H + m_C) v_{f,y} \qquad \textbf{(16)}$$

The two components of the final velocity are isolated in the two formulas in Eq. [26], respectively. For the x–component we find:

$$v_{f,x} = \frac{m_H v_H \cos\theta + m_C v_C}{m_H + m_C} = +7.9 \frac{m}{s} \qquad \textbf{(17)}$$

and for the y–component:

$$v_{f,y} = -\frac{m_H v_H \sin\theta}{m_H + m_C} = -12.5 \frac{m}{s} \qquad \textbf{(18)}$$

Thus, the speed of the hawk after catching the crow is:

$$v_{final} = \sqrt{v_{f,x}^2 + v_{f,y}^2} = 14.8 \frac{m}{s} \qquad \textbf{(19)}$$

which corresponds to a sudden reduction in speed of more than 25 % for the hawk.

Solution to part (b): Fig. 4.31 shows the final velocity in the xy–plane. We define the angle φ with the positive x–axis. Note that the velocity is a vector in the fourth quadrant and that φ is shown in a clockwise rotation from the x–axis. This orientation should also result when calculating the angle based on the velocity components:

$$\tan\varphi = \frac{v_{f,y}}{v_{f,x}} = \frac{-12.5 \frac{m}{s}}{7.9 \frac{m}{s}} = -1.58$$
$$\Rightarrow \quad \varphi = -57.7^0 \qquad \textbf{(20)}$$

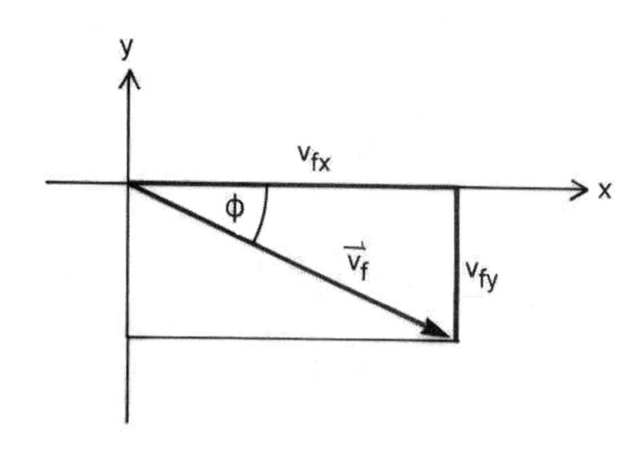

Figure 4.31

The negative sign in Eq. [20] indicates that the angle is measured from the x–axis in clockwise rotation. Thus, the change in angle is $70^0 - 57.7^0 = 12.3^0$. Hawks do not just crash into mid–sized birds in air while hunting. The problem is not so much the change in angle as calculated in Eq. [20] but the sudden change in speed as found in Eq. [19]. Note that any modification to the approach of hunting that reduces the impact ultimately requires an adjustment in the speed of the bird of prey. For the bird of prey to be successful, this change must come as late as possible since it allows the prey extra time to escape.

Problem 4.5
Ruby–throated hummingbirds can be watched hovering at feeders throughout eastern North America. These birds have a wingspan of 11 cm and a mass of just 3.0 g. Using the momentum disc theory, to what speed does the bird have to accelerate air downward in order to maintain its hovering position? Use $\rho_{air} = 1.2$ kg/m³.

Solution: We use the momentum disc theory to find the required speed of air in Eq. [4.29]. Even though this formula was derived for hovering insects, it does apply to the hummingbird as they can hover. In the problem text, the mass of the bird and the density of air are given. The area of the stroke plane is circular with the diameter given by the wingspan as $A = \pi \cdot r^2 = \pi \cdot d^2/4$ with d = 11 cm. We substitute these values in Eq. [4.30]:

$$v = \frac{1}{d}\sqrt{\frac{4 \cdot m \cdot g}{\rho \cdot \pi}}$$
$$= \frac{1}{0.11\ m}\sqrt{\frac{4\left(3.0 \times 10^{-3}\ kg\right)\left(9.8 \frac{m}{s^2}\right)}{\pi \cdot 1.2 \frac{kg}{m^3}}} \qquad \textbf{(21)}$$

which yields $v = 1.6$ m/s. Note that we calculated a speed of air of 2.2 m/s for the bumble bee; thus, the humming bird is slightly better adapted to hovering if we interpret the value of v as a measure of evolutionary adaptation.

Analysing Eq. [21] a bit further and studying the question why so few bird species have mastered hovering flight is very interesting. Intuitively, we believe that it is the small size of the hummingbird that allows it to hover. However, Eq. [21] rules out an advantage for the hummingbird with respect to the momentum disc theory! Here is why: there are two parameters in Eq. [21] that vary between various bird species: the bird's mass and its wingspan. These are not independent parameters because there are common basic design features among all birds that can fly. These were noted already in Problem 1.5. We found that a square–root relation holds between the wingspan d and the mass m of a bird: $d \propto m^{1/2}$. Substituting this relation into Eq. [21] yields:

$$v = \frac{1}{d}\sqrt{\frac{4 \cdot m \cdot g}{\rho \cdot \pi}} \propto \frac{\sqrt{m}}{d} \propto \frac{d}{d} = const \qquad \textbf{(22)}$$

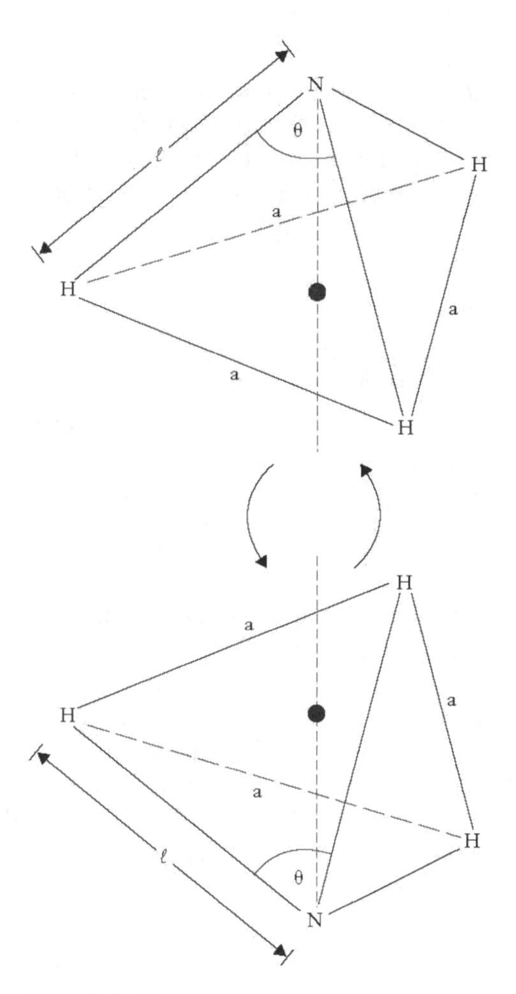

FIGURE 4.22

It is still correct that a Canada goose cannot hover. We will discover why in Chapter 6.

Problem 4.7

An object of mass $m = 3.0$ kg makes a perfectly inelastic collision with a second object that is initially at rest. The combined object moves after the collision with a speed equal to one–third of the object that was initially moving. What is the mass of the object that was initially at rest?

Solution: Use conservation of momentum. Eliminate the speed v and solve for the unknown mass M; $M = 6.0$ kg.

Problem 4.9

In the ammonia molecule NH_3 the three hydrogen atoms are located in a plane forming an equilateral triangle with side length a as shown in Fig. 4.22. The nitrogen atom oscillates 24 billion times per second up and down along a line that intersects the plane of the hydrogen atoms at the centre of mass of the three hydrogen atoms (solid circles in Fig. 4.22).

(a) Calculate the length a in Fig. 4.22, using for the N–H bond length $l = 0.1014$ nm, and for the HNH–bond angle $\theta = 106.8^0$.

(b) Calculate the distance between the centre of mass of the three hydrogen atoms and any one of the hydrogen atoms.

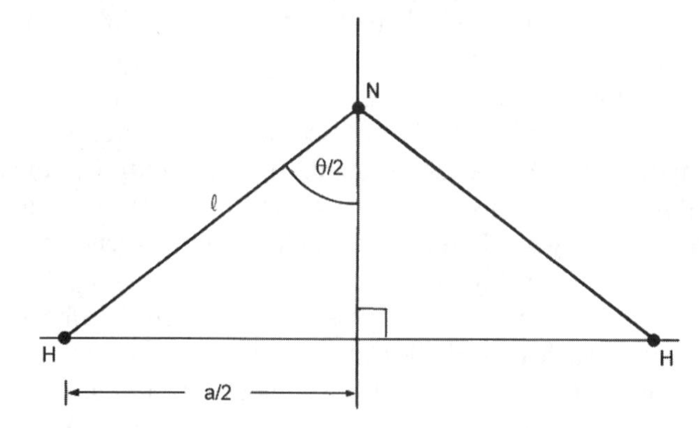

Figure 4.32

Solution to part (a): Fig. 4.32 shows the nitrogen atom and two of the hydrogen atoms of the ammonia molecule drawn in a plane. The symmetry of the molecule allows us to construct a right triangle by enclosing a line dividing the molecule in two equal halves. The angle at the nitrogen atom is $\theta/2 = 53.4^0$ and the length of the side opposite to the nitrogen atom is $a/2$. From trigonometry we know:

$$\sin\left(\frac{\theta}{2}\right) = \frac{a/2}{l}$$

$$\Rightarrow \quad a = 2 \cdot (0.1014\ nm)\ \sin(53.4^0) \tag{23}$$

which yields $a = 0.163$ nm.

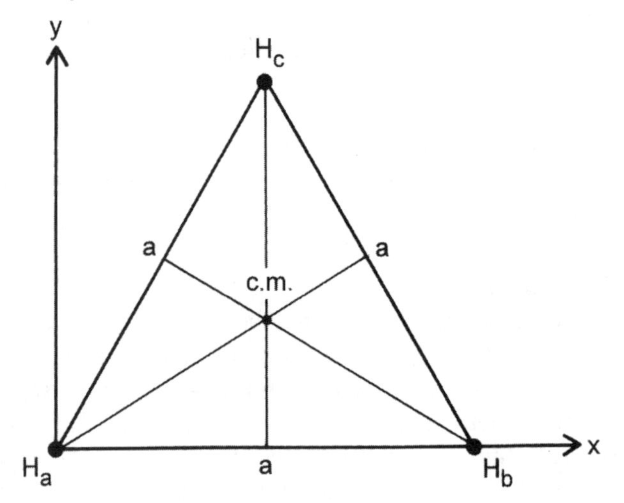

Figure 4.33

Solution to part (b): The centre of mass position of the three hydrogen atoms in the ammonia molecule is indicated as a dot in Fig. 4.22. Fig. 4.33 shows the position of the hydrogen atoms in their common plane. We identify this plane as the *xy*–plane and choose the two axes as shown in the figure.

From Fig. 4.33, the Cartesian coordinates of the hydrogen atoms are determined. For this purpose, the three atoms are distinguished by indices a, b, and c. Note that the indices do not imply a chemical difference between the three hydrogen atoms.

The coordinate system in Fig. 4.33 has been chosen such that H_a lies at the origin. Therefore, we know that $H_a = (0,0)$. We further choose the *x*–axis such that it coincides with the line connecting H_a and H_b. With the distance *a* calculated in part (a), the position of H_b is established as $H_b = (a, 0)$. Thus, only the position of H_c requires some reasoning.

We know that the symmetry of the molecule requires that the three hydrogen atoms form an equilateral triangle (all three side lengths are equal to *a* as shown in Fig. 4.33). Thus, due to symmetry, the *x*–component of the position of H_c lies half–way between the *x*–components of H_a and H_b, i.e., $H_{c,x} = a/2$. With this information, the *y*–component of the position of H_c is calculated using the right triangle shown in Fig. 4.34. The figure shows the left half of the triangle in Fig. 4.32,

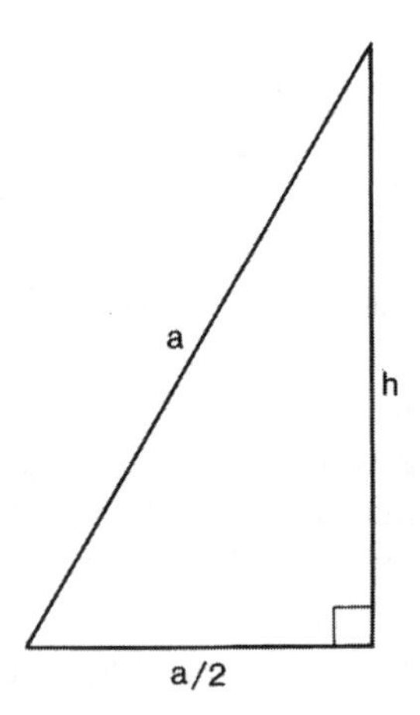

Figure 4.34

with *h* the side adjacent to the right angle and H_c. To find *h*, we use the Pythagorean theorem:

$$a^2 = \left(\frac{a}{2}\right)^2 + h^2$$

$$\Rightarrow \quad h = \frac{\sqrt{3}}{2}\, a \tag{24}$$

The position of atom H_c is $H_c = (a/2, a\sqrt{3}/2)$.

Next the coordinates of the centre of mass position are determined. We use Eq. [4.37] for the *x*–component of the centre of mass position, and an analogous equation for its *y*–component. The mass of the three hydrogen atoms is the same and is labelled *m*:

$$x_{cm} = \frac{0\,m + a\,m + \frac{a}{2}\,m}{m + m + m} = \frac{\frac{3}{2}a\,m}{3\,m} = \frac{a}{2}$$

$$y_{cm} = \frac{0\,m + 0\,m + \frac{\sqrt{3}}{2}a\,m}{m + m + m} = \frac{a}{2\sqrt{3}} \tag{25}$$

In the last step, the distance between one of the hydrogen atoms and the centre of mass position is calculated. The symmetry of the molecule allows us to pick any one of the hydrogen atoms as their distances to the *cm*–position are equal. We pick H_a because that reduces the calculation to determining the distance of the *cm*–position from the origin. Using the data in Eq. [25] we find for the distance from H_a to the *cm*–position:

$$|\boldsymbol{H_a\, cm}| = \sqrt{\left(\frac{a}{2}\right)^2 + \left(\frac{a}{2\sqrt{3}}\right)^2} = \frac{a}{\sqrt{3}} \tag{26}$$

Substituting the value for a from the problem part (a), we find that the distance from each hydrogen atom to the centre of mass is 0.094 nm.

Problem 4.11

(a) An object moves with initial speed $v_0 = 10.0$ m/s on a horizontal surface. It slides for a distance of 20.0 m before it comes to rest as illustrated in Fig. 4.23(a). Determine the coefficient of kinetic friction between object and surface.

(b) How long does the object move until it comes to rest?

(c) With what speed does the object move after the same time if the surface is tilted to an angle of $\theta = 10^0$ with the horizontal and the object is sliding downhill, as illustrated in Fig. 4.23(b)?

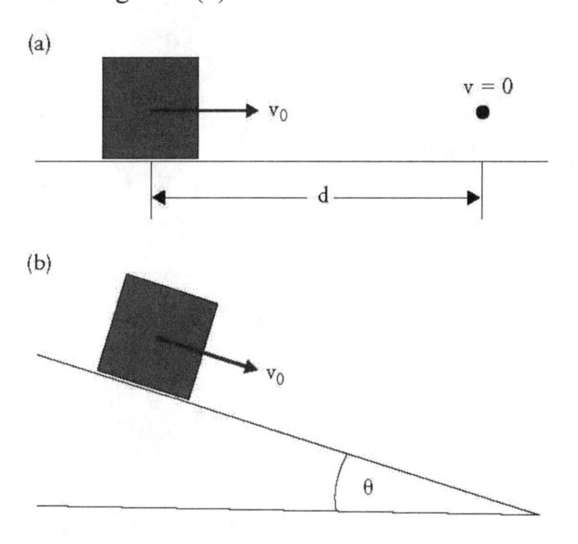

FIGURE 4.23

Solution to part (a): Fig. 4.35(a) shows the free–body diagram for the object, Fig. 4.35(b) illustrates the main kinematic properties of the system. Weight and normal force balance each other in the vertical direction. Friction is the only horizontal force. It provides therefore an acceleration in the direction opposite to the motion. The direction of the initial motion is defined by the velocity vector $\mathbf{v}_0$. We choose the x–direction in the direction of $\mathbf{v}_0$. Since this is the only direction of interest, no vector algebra is needed as long as we choose the signs of the kinematic components carefully.

Newton's second law along the x–axis is written with Fig. 4.35(a):

$$F_{net} = -\mathscr{F}_k = m \cdot a$$
$$with \ |\mathscr{F}_k| = \mu_k \cdot N = \mu_k m \cdot g \quad \textbf{(27)}$$

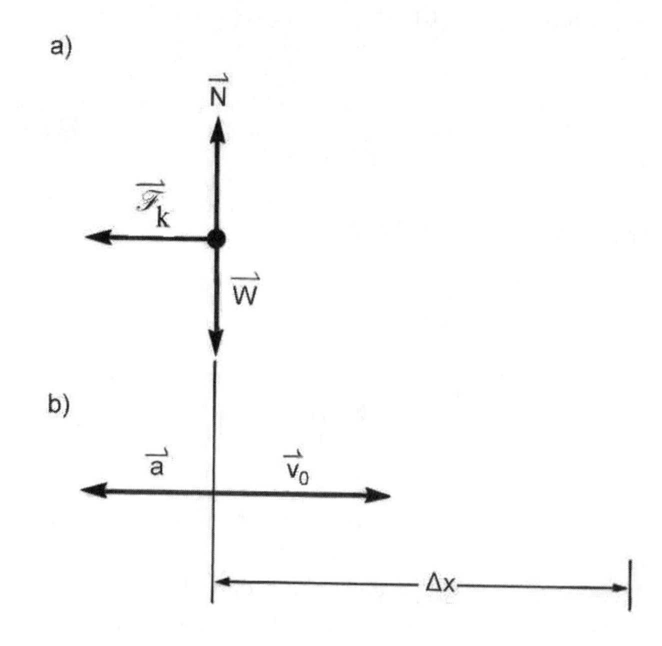

Figure 4.35

in which $\mathscr{F}_k$ is the kinetic friction force, μ_k is the coefficient of kinetic friction, **N** is the normal force and **W** = $m \cdot \mathbf{g}$ the weight of the object. Note that the net force points in the negative x–direction. Substituting the second formula in Eq. [27] in the first we obtain:

$$-\mathscr{F}_k = -\mu_k \cdot m \cdot g = m \cdot a \quad \textbf{(28)}$$

which yields:

$$a = -\mu_k \cdot g \quad \textbf{(29)}$$

Now we use the kinematic relations to relate displacement, initial speed and acceleration:

$$x - x_0 = v_0 \cdot t + \frac{1}{2} a \cdot t^2$$
$$v = v_0 + a \cdot t \quad \textbf{(30)}$$

in which $x - x_0$ is the displacement $\Delta x = 20$ m and $v_0 = 10$ m/s is the initial speed. We eliminate the time t in Eq. [30]. Using Δx we identify t as the time the object requires to come to rest, this time is eliminated as we do not require it in the calculations. Using $t = -v_0/a$ in the first formula of Eq. [30] we find:

$$\Delta x = v_0\left(-\frac{v_0}{a}\right) + \frac{1}{2}a\left(-\frac{v_0}{a}\right)^2 = -\frac{v_0^2}{2a} \quad \textbf{(31)}$$

Substituting Eq. [29] in Eq. [31] we obtain:

$$\Delta x = \frac{v_0^2}{2 \cdot \mu_k \cdot g} \qquad \textbf{(32)}$$

We calculate μ_k from Eq. [32]:

$$\mu_k = \frac{v_0^2}{2 \cdot \Delta x \cdot g} = \frac{\left(10 \frac{m}{s}\right)^2}{2\,(20\, m)\left(9.8 \frac{m}{s^2}\right)} = 0.26 \qquad \textbf{(33)}$$

Note that the coefficient of kinetic friction has no units.

Solution to part (b): We use the second formula in Eq. [30] in the form $t = -v_0/a$ with $a = -\mu_k \cdot g$ from Eq. [29]:

$$t = \frac{v_0}{\mu_k \cdot g} = \frac{10 \frac{m}{s}}{0.26 \cdot 9.8 \frac{m}{s^2}} = 4.0\, s \qquad \textbf{(34)}$$

The object slows down to rest within 4 seconds.

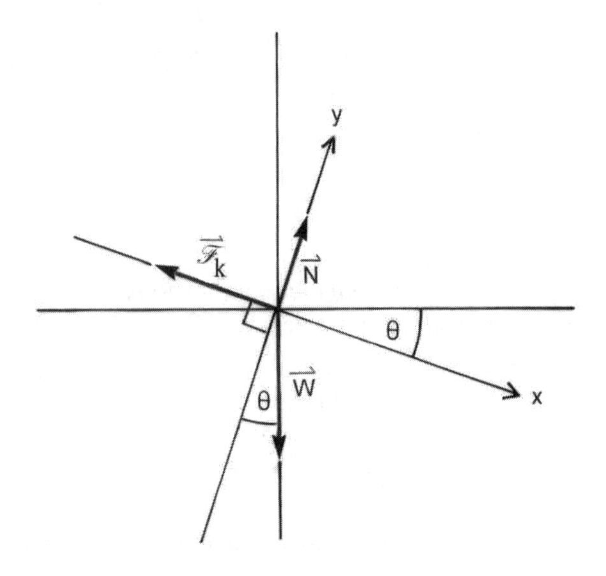

Figure 4.36

Solution to part (c): The free–body diagram for the object on the inclined surface is shown in Fig. 4.36 with $\theta = 10^0$. Three forces act on the object: its weight **W** vertically downward, the normal force **N** perpendicular to the surfaces, and the frictional force $\mathscr{F}_k$ against the motion of the object, i.e., upward along the inclined surface. We choose the *x*–axis parallel to the inclined surface and the *y*–axis perpendicular. Thus, Newton's first law applies along the *y*–axis and Newton's second law along the *x*–axis:

x–direction:

$$W \sin\theta - \mathscr{F}_k = m \cdot a_x$$

y–direction:

$$N - W \cos\theta = 0 \qquad \textbf{(35)}$$

The second formula in Eq. [35] yields $N = W\cos\theta$ which allows us to rewrite the frictional force in the first formula of Eq. [35] in the form $\mathscr{F}_k = \mu_k \cdot N = \mu_k \cdot W\cos\theta$. With this substitution the first formula in Eq. [35] reads:

$$W \sin\theta - \mu_k \cdot W \cos\theta = m \cdot a_x \qquad \textbf{(36)}$$

Substituting $W = m \cdot g$ in Eq. [36] yields the acceleration along the inclined surface:

$$\begin{aligned} a_x &= (\sin\theta - \mu_k \cos\theta)\, g \\ &= (\sin 10^0 - 0.26 \cos 10^0)\; 9.8 \frac{m}{s^2} \\ &= -0.81 \frac{m}{s^2} \end{aligned} \qquad \textbf{(37)}$$

In four seconds, this acceleration changes an initial speed of 10 m/s to:

$$\begin{aligned} v &= v_0 + a_x \cdot t \\ &= 10 \frac{m}{s} - \left(0.81 \frac{m}{s^2}\right)(4\, s) = 6.76 \frac{m}{s} \end{aligned} \qquad \textbf{(38)}$$

which is about 2/3 of the initial speed.

Problem 4.13

A standard man rides on a bicycle. Resistance against the forward motion is due to air resistance, friction of the tires on the road and the metal–on–metal sliding in the lubricated axles. Combining the latter two effects as a velocity–independent friction force $\mathscr{F}$, determine the speed above which air resistance limits the cyclist's motion. $\mathscr{F}$ is determined from the fact that the cyclist rolls with constant speed downhill on a road that has a 1 % slope. Use $\rho_{air} = 1.2\ kg/m^3$ and a cross–sectional area $A = 0.5\ m^2$

for the bicycle/standard man system in the direction of motion. Use $D = 0.5$ for the drag coefficient.

Solution: We use Eq. [4.50] for the air resistance:

$$F_{drag} = \frac{D}{2}\rho \cdot A \cdot v_{system}^2 \quad \textbf{(39)}$$

and Eq. [4.56] for the combined friction effect:

$$\mathcal{F}_k = \mu_k \cdot N \quad \textbf{(40)}$$

The information given in the problem text allows us to circumvent the coefficient for kinetic friction. A 1 % slope means that the road drops by 1 m for every 100 m length, i.e., the road has an angle of θ with $\tan\theta = 1/100$. At this small angle we use the approximation that $\tan\theta = \sin\theta = \theta$ and $\cos\theta = 1$. Fig. 4.36 can be used again in this case, this time with $N = W$ (because $\cos\theta = 1$) and $\mathcal{F}_k = W\sin\theta = 0.01 \cdot W$ due to the mechanical equilibrium that applies along the direction of the inclined surface.

We set Eqs. [39] and [40] equal to find the speed at which the two resistance forces match each other:

$$F_{drag} = \frac{D}{2}\rho \cdot A \cdot v_{system}^2 = 0.01 \cdot W = \mathcal{F}_k \quad \textbf{(41)}$$

This yields the speed of the system (if we neglect the mass of the bicycle and use 70 kg for the standard man):

$$v_{system} = \sqrt{\frac{2 \cdot 0.01 \cdot W}{D \cdot \rho \cdot A}} = \sqrt{\frac{0.02\,(70\,kg)\left(9.8\,\frac{m}{s^2}\right)}{0.5\left(1.2\,\frac{kg}{m^3}\right)(0.5\,m^2)}} = 6.76\,\frac{m}{s} = 24.3\,\frac{km}{h} \quad \textbf{(42)}$$

At speeds below 24 km/h the friction on the road and in the axles dominates, at speeds above 25 km/h the air resistance dominates due to the v^2 term in the air drag formula, Eq. [39]. Note that this is the relative speed between the bicycle and the air. Thus, a 25 km/h head wind will dominate the cyclist's motion even if the bicycle is moving very slowly with respect to the ground.

Problem 4.15

An object of mass $m = 20$ kg is initially at rest on a horizontal surface. It requires a horizontal force $F = 75$ N to set in motion. However, once in motion, only a horizontal force of 60 N is required to keep it moving with a constant speed. Find the coefficients of static and kinetic friction in this case.

Solution: The weight of the object is 196 N, and for static equilibrium the normal force exerted by the hori-zontal surface on the object must also be 196 N. This leads to:

$$\mu_s = \frac{\mathcal{F}_s}{N} = \frac{75\,N}{196\,N} = 0.38 \quad \textbf{(43)}$$

Similarly, $\mu_k = 0.31$.

Problem 4.17

In Fig. 4.25, the coefficient of static friction is 0.3 between an object of mass $m = 3.0$ kg and a surface, which is inclined by $\theta = 35^0$ with the horizontal. What is the minimum magnitude of force **F** that must be applied to the object perpendicular to the inclined surface to prevent the object from sliding downward?

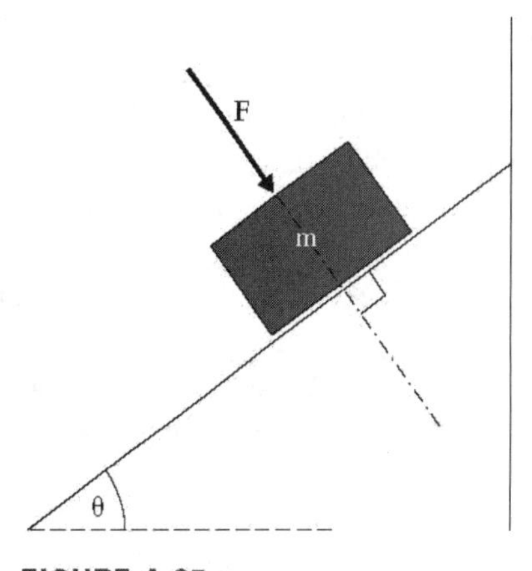

FIGURE 4.25

Solution: Set up a coordinate system such that positive x is parallel to the ramp surface, increasing as the block moves up the ramp. The y-axis is perpendicular to the ramp surface and increases upwards and to the left. In the x direction, this gives us $\mathcal{F}_S - W\sin\theta = 0$. In the y direction, $N - W\cos\theta - F = 0$. Solving for N and using $\mathcal{F}_S = \mu_s \cdot N$ leads to:

$$F = W\frac{\sin\theta - \mu_s \cdot \cos\theta}{\mu_s} = 32.1\,N \quad \textbf{(44)}$$

CHAPTER FIVE

Kinesiology: The action of Forces at Joints

MULTIPLE CHOICE AND CONCEPTUAL QUESTIONS

Question 5.1

Note: In certain professional admission tests, some multiple choice questions are grouped with a common text called a *passage*. We illustrate this approach with the first four questions in this section.

Passage: We study a standard man bending forward to lift an object off the ground, as illustrated in Fig. 5.43(a). The fulcrum lies at the lower back and the spinal column is the lever arm. We assume that the person is bending forward with the back horizontal and that the forearms are stretched downward. For this case we obtain the balance of torque diagram, shown in Fig. 5.43(b), with four forces acting on the spinal column:

- the weight of the torso, $\mathbf{W}_2$, acting at the centre of mass located halfway between the two ends of the spinal column,
- the weight of the object that is lifted and combines with the weight of the arms and hands as $\mathbf{W}_1$,
- the tension **T** exerted on the spinal column by the back muscle responsible for the lifting. We assume this force is applied at a distance $d = 2 \cdot L/3$ from the fulcrum at an angle θ above the horizontal.
- A compressive force, **R**, exerted on the fulcrum in the lower back. Let β be the angle between **R** and the horizontal.

Question (i): An author reporting on this case may opt to exclude the force **R** from Fig. 5.43(b). Why would the author have done so? (A) The author studies a problem of rotational equilibrium and anticipates that this force will not be required when writing the torque formula. (B) Different from a free–body diagram, we may eliminate up to all but one force from a diagram such as Fig. 5.43(b). (C) The physiological knowledge about the magnitude of the force **R** and its angle β is uncertain; thus, omitting the force in Fig. 5.43(b) would eliminate a possible source of error. (D) For the person in the posture shown in Fig. 5.43(a), the force **R** indeed doesn't exist, i.e., R = 0. (E) Fig. 5.43(b) oversimplifies the problem as the force **R** acts in a direction out of the plane of the paper.

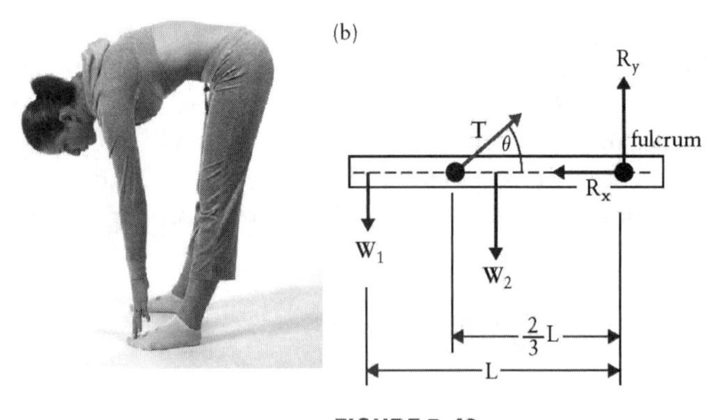

FIGURE 5.43

Question (ii): Which of the following four formulas is the proper torque equation for the problem illustrated in Fig. 5.43? Note that the term "proper" includes the use of the generally accepted sign convention for torque as introduced in the textbook.

(*A*)
$$\tau_{net} = L\,W_1 + \frac{L}{2}W_2 + \frac{2}{3}L\,T\sin\theta = 0$$

(*B*)
$$\tau_{net} = L\,W_1 + \frac{L}{2}W_2 - \frac{2}{3}L\,T\sin\theta = 0$$

(*C*)
$$\tau_{net} = -L\,W_1 - \frac{L}{2}W_2 + \frac{2}{3}L\,T\sin\theta = 0 \qquad \textbf{(1)}$$

(*D*)
$$\tau_{net} = -L\,W_1 - \frac{L}{2}W_2 - \frac{2}{3}L\,T\sin\theta = 0$$

Question (iii): Assume that Fig. 5.43(a) were instead drawn such that the standard man is shown from the opposite side, i.e., bending down toward the right. Correspondingly, Fig. 5.43(b) would be drawn with the left and right ends flipped horizontally. For this modified display, which of the four choices in Eq. [1] would now be the proper torque equation?

Question (iv): If you answered the two previous questions choosing the same formula in Eq. [1], then skip this question. If you have chosen different answers for the two previous questions, then choose the statement below that best describes the consequences of your choices: (A) We need two different formulas because the two cases are physically different. (B) The two formulas we have chosen are mathematically equivalent (differ only due to the sign convention). This is correct because the two cases are physically the same. (C) The two cases should be the same, but the two formulas are indeed different. Thus, there must be something wrong with the introduced sign convention. (D) The two cases differ physically. That the two formulas are mathematically equivalent is accidental.

Answer to Question (i): (A).

Answer to Question (ii): (B). Remember to follow the sign convention discussed in the text.

Answer to Question (iii): (C).

Answer to Question (iv): (B). The sign convention is designed such that the mathematical description will produce identical results regardless of point of view.

Question 5.3
The definition of torque contains the magnitude of the force **F** acting on a rigid object with a fixed axis, the magnitude of the vector **r** between the fulcrum and the point where the force is applied, and the angle φ between the force vector **F** and position vector **r**. Which of the following statements about torque is wrong? (A) Torque is linearly proportional to the magnitude of the force **F**. (B) Torque is linearly proportional to the magnitude of the vector **r**. (C) Torque is linearly proportional to the angle φ. (D) Torque can be positive or negative depending on the angle φ. (E) The force **F** can be applied to the rigid object such that the resulting torque is zero.

Answer: (C). $\tau \propto \sin\varphi$ is not a linear function in φ.

Question 5.5
A rod is 7 metres long and is pivoted at a point 2.0 m from the left end. A force of magnitude 50 N acts downward at the left end and a force of magnitude 200 N at the right end. At what distance from the pivot point must a third upward–directed force of magnitude 300 N be placed to establish rotational equilibrium? Neglect the weight of the rod. (A) 1.0 m, (B) 2.0 m, (C) 3.0 m, (D) 4.0 m, (E) none of the first four answers is correct.

Answer: (C).

Question 5.7
What is the major difference between
(a) hinge and pivot joints, and
(b) ellipsoid and saddle joints?

Answer to part (a): The axis of rotation through the joint is perpendicular to the axis of the two adjacent bones for the hinge joint, while both axes are parallel for the pivot joint. Thus, the difference is the direction of the resulting rotation.

Answer to part (b): The two axes of rotation are identical for the ellipsoid and saddle joints, however, one of the actual motions is due to the bone farthest from the trunk in the case of the saddle joint. This is more complicated to arrange, and thus this type of joint is seldom seen in nature.

Question 5.9
Why does holding a long pole help a tightrope walker stay balanced?

Answer: The centre of mass of the person must always be directly above the rope. Small corrections with a long pole are effective because the ends of the pole produce a large torque.

Question 5.11
A typical mass distribution of a tree is 60% for the trunk, 20% for the branches, and 20% for the root system. Despite the great heights that trees can grow, most of them have shallow root systems not reaching deeper than about 60 cm below the surface. Considering once more Fig. 4.16, determine why the root system may not be a sufficient anchor to protect against strong winds, as noted in Table 5.1, which defines the Beaufort scale. Can

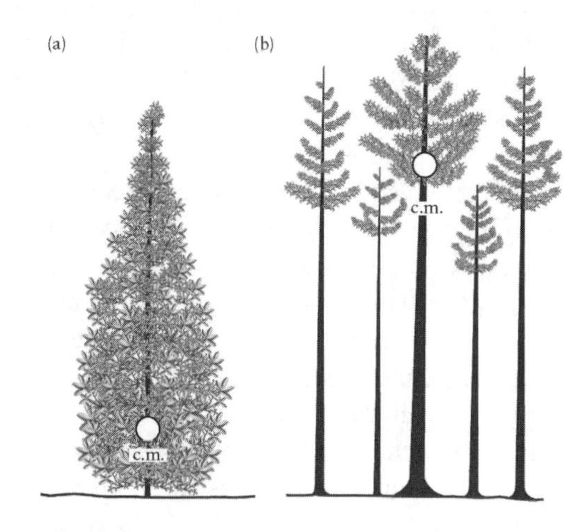

FIGURE 4.16

Table 5.1

Beaufort No.	Wind speed (km/h)	Name
Effect on trees		
0 – 2	0 – 11	calm to light breeze
leaves rustle		
3 – 4	12 – 28	gentle to moderate breeze
leaves and small twigs in continuous motion, small branches move		
5 – 6	29 – 49	fresh to strong breeze
small–leaved trees sway, large branches move		
7	50 – 61	moderate gale
whole trees sway		
8 – 9	62 – 88	fresh to strong gale
twigs break off		
10 – 11	89 – 117	storm to violent storm
trees uprooted		
12	> 117	hurricane
widespread damage		

you think of a reason why these shallow root systems evolved? Trees appeared in the fossil record as early as 370 million years ago, with most modern species emerging during the last 65 million years. Trees should therefore not display serious misadaptations.

Answer: The primary purpose of the root system is to supply the tree with water and minerals. Water is mostly accessible at shallow depths, and trees developed therefore shallow root systems that may reach twice as far outward as the tree top. This makes sense when you watch a coniferous tree in the rain: water reaches the ground mostly beyond the "drip–line" which is the enveloping line of the vertical projection of the tree onto the ground. The root system serves a second purpose providing stability against torques due to horizontal forces acting at the tree's centre of mass. The fulcrum is located where the trunk enters the ground. The torque on the tree above ground must be balanced by a counter–torque based on the horizontal force the ground exerts on the tree's short lever arm below ground. A deep root would serve this purpose best.

When two competing concepts create an evolutionary problem, nature find adaptations to address the issue or the affected species goes extinct. Trees developed two alternate ways to address this issue, both illustrated in Fig. 4.16: either they grow in forests together where the horizontal forces acting on individual trees during storms are diminished through mutual screening, or they have branches all the way down to the ground. This lowers the centre of mass as the total mass contribution of the branches is a notable 20%, or 1/3 of the mass of the trunk. For coniferous trees it also allows the tree to buffer storm–related forces as branches sway in the wind while the thrust on the trunk is diminished. Branches provide coniferous trees therefore with a protection comparable to an air–bag system in a car. Trees get uprooted, but Table 5.1 illustrates that very strong winds are required for this to happen.

Question 5.13

Fig. 5.47 shows a one–bottle wine holder. What do we know about the centre of mass of the system bottle and wine holder if it is in rotational equilibrium, i.e., doesn't fall over?

Figure 5.47

Answer: The centre of mass of the bottle plus wine holder system must be located directly above the support point on the table because this leads to a net torque of zero.

ANALYTICAL PROBLEMS

Problem 5.1

If the torque required to loosen a nut has a magnitude of $\tau = 40\ \mathrm{N \cdot m}$, what minimum force must be exerted at the end of a 30-cm-long wrench?

Solution: $F_{min} = \tau / d = (40\ \mathrm{N \cdot m}) / 0.3\ \mathrm{m} = 133\ \mathrm{N}$.

Problem 5.3

A steel band of a brace exerts an external force of magnitude $F_{ext} = 40$ N on a tooth. The tooth is shown in Fig. 5.44, with point B a distance 1.3 cm above point A, which is the fulcrum. The angle between the tooth and the external force is $\theta = 40^0$. What is the torque on the root of the tooth about point A?

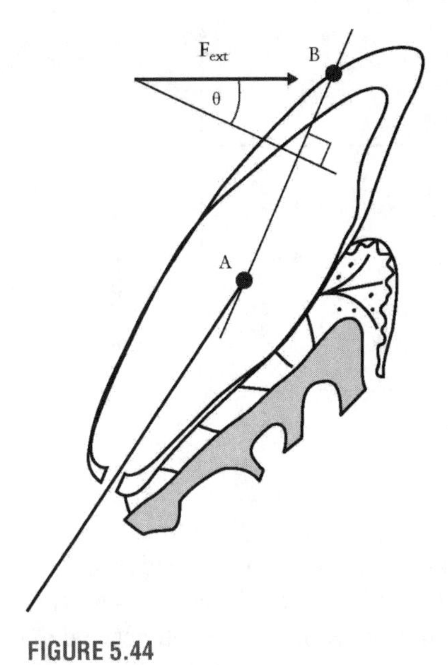

FIGURE 5.44

Solution: This is an application of the torque definition for an extended object, but contains a geometrical twist: there is only one force to be considered, but the angle θ is not the angle we need to analyse Eq. [5.19] for the torque. To use that equation, the angle between the exerted force, $\mathbf{F}_{ext}$, and the lever arm vector, that is **AB** = **r**, is needed. This is illustrated in Fig. 5.62, where the angle between both vectors is defined as φ. Confirm that the angle θ as shown in Fig. 5.44 and the angle θ as shown in Fig. 5.62 are indeed the same. Remember that vectors can be moved as long as their length and direction are kept the same. In this case, we can move vector $\mathbf{F}_{ext}$ while keeping the same angle between $\mathbf{F}_{ext}$ and **AB**. In particular, we can slide $\mathbf{F}_{ext}$ to the right in Fig. 5.44 so that its tail end is located at point B. The angles in this problem are related in the following form:

$$\varphi = 180^0 - (90^0 - \theta) = 90^0 + \theta \qquad \textbf{(2)}$$

With all necessary parameters defined, we use Eq. [5.19] to calculate the torque:

$$\tau = l \cdot F_{ext} \sin\varphi$$
$$= (1.3 \times 10^{-2}\ m)\ (40\ N) \sin 130^0 = 0.4\ N \cdot m \qquad \textbf{(3)}$$

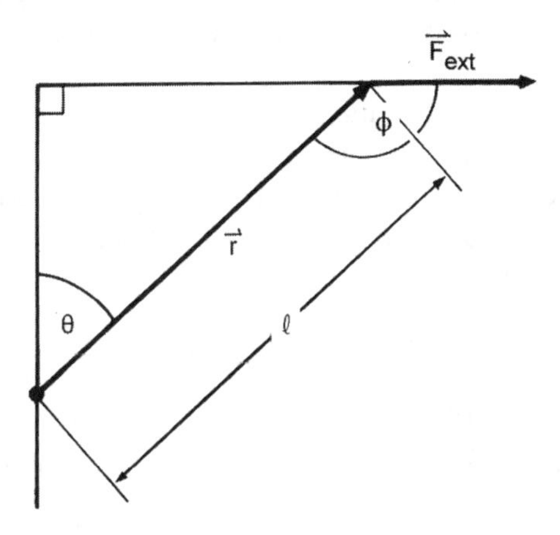

Figure 5.62

Problem 5.5

Fig. 5.48 shows a light of mass m = 20 kg supported at the end of a horizontal bar of negligible mass that is hinged to a pole. A cable at an angle of 30^0 with the bar helps to support the light. Find

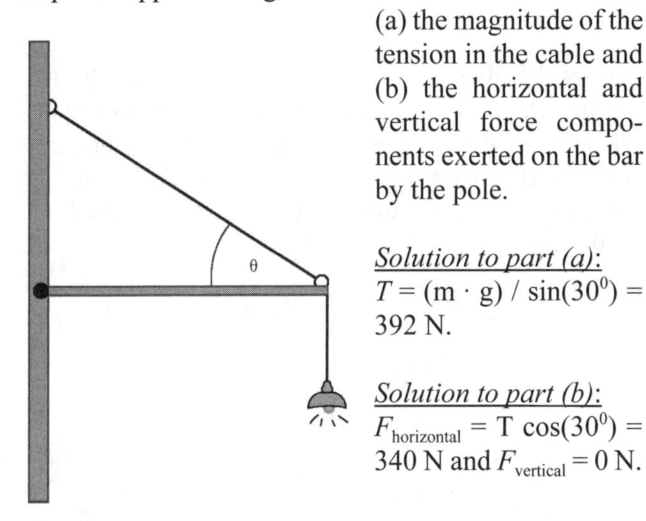

FIGURE 5.48

(a) the magnitude of the tension in the cable and
(b) the horizontal and vertical force components exerted on the bar by the pole.

Solution to part (a):
$T = (m \cdot g) / \sin(30^0)$ = 392 N.

Solution to part (b):
$F_{horizontal} = T \cos(30^0)$ = 340 N and $F_{vertical}$ = 0 N.

Problem 5.7

Fig. 5.49 shows a uniform boom of mass m = 120 kg supported by a cable perpendicular to the boom. The boom is hinged at the bottom and an object of mass M= 200 kg hangs from its top. Find the tension in the massless cable and the components of the force exerted on the boom at the hinge.

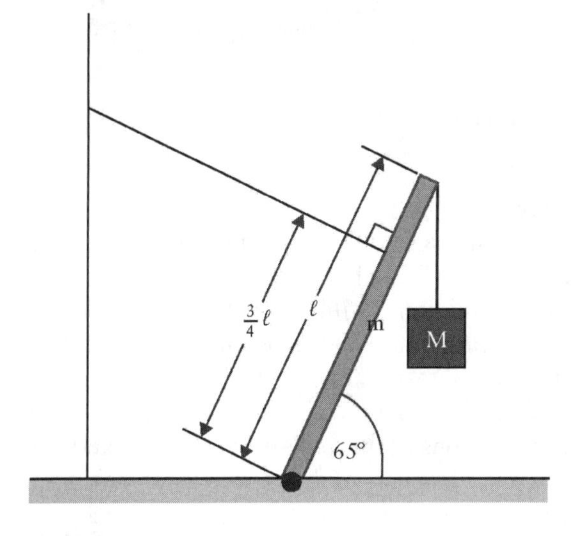

FIGURE 5.49

Solution: First, consider the torques about the point where the boom contact the ground. This gives

$$T = \frac{M \cdot g \cdot \cos\theta \cdot l + m \cdot g \cdot \cos\theta \cdot \frac{l}{2}}{\frac{3}{4} l} \quad \textbf{(4)}$$

Notice that there is no dependence on the length of the beam, which is convenient because we do not have a value for that length. Solving Eq. [4], we find that T = 1435 N. Balancing forces in the horizontal direction, we find that $F_{horizontal} = T \sin\theta = 1300$ N. Balancing the forces in the vertical, we find:

$$F_{vertical} = M \cdot g + m \cdot g - T \cos\theta = 2530\ N \quad \textbf{(5)}$$

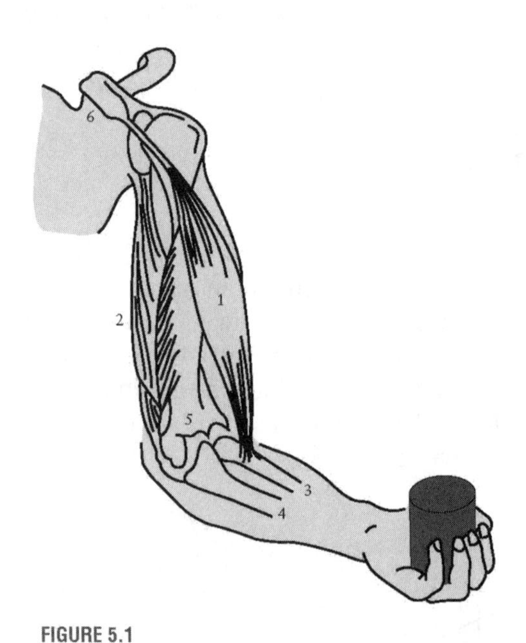

FIGURE 5.1

Problem 5.9

A standard man holds the upper arm vertical and the lower arm horizontal with an object of mass M = 6 kg resting on the hand as illustrated in Fig. 5.1. The mass of the lower arm and hand is one half of the mass of the entire arm. Shown in Fig. 5.51 is the arrangement of the four forces acting on the lower arm that we include in the calculations: (I) the external force $\mathbf{F}_{ext}$, exerted by the bones and ligaments of the upper arm at the elbow (fulcrum), (II) the tension **T**, exerted by the biceps, (III) a force **F** due to the weight $\mathbf{W}_M$ of the object of mass M, (IV) the weight $\mathbf{W}_F$ of the lower arm. The points along the lower arm, at which the forces act, are identified in Fig. 5.51: l_1 = 4 cm, l_2 = 15 cm and l_3 = 40 cm.

(a) Calculate the vertical component of the force $\mathbf{F}_{ext}$, and

(b) calculate the vertical component of the tension **T**.

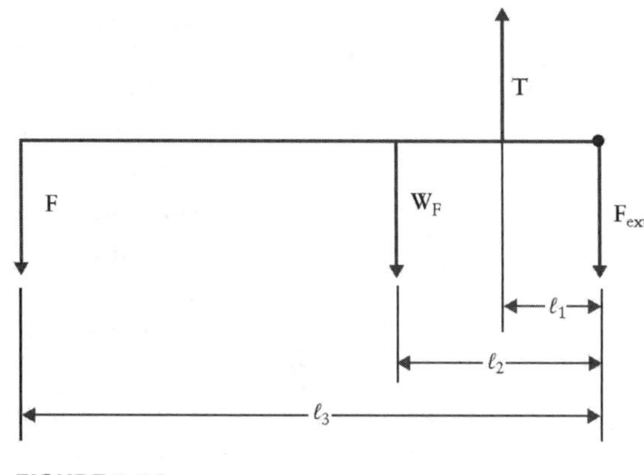

FIGURE 5.51

Solution to part (a): Fig. 5.51 provides already a balance of torque diagram, while Fig. 5.1 may serve as a sketch from which Fig. 5.51 is derived. When sketches and diagrams are already given, you want to take extra care to familiarize yourself with the choices made since they may differ from what you might have done yourself. You notice that the free–body diagram is not shown. It should be drawn before proceeding.

Since all the forces in Fig. 5.51 act along a vertical direction which we define as the y–axis, only two formulas, Newton's first law for the y–components of the forces (condition 2) and the torque equilibrium equation (condition 3) are needed:

$$\begin{aligned} &condition\ 2: \quad T - F_{ext} - W_F - F = 0 \\ &condition\ 3: \quad -T l_1 + W_F l_2 + F l_3 = 0 \end{aligned} \quad \textbf{(6)}$$

For condition 3 note that the fulcrum is at the right end of the bar in Fig. 5.51, where the force $\mathbf{F}_{ext}$ acts on the lower arm. Note that the biceps tendon in a real arm does not act perfectly perpendicular to the lower arm. Thus, assuming that the tension **T** is directed in the y–direction is a simplifying assumption. That this assumption simplifies the calculations becomes evident when we later get to examples where some forces do not act perpendicular to the bar.

We have two formulas in Eq. [6] with two unknown variables, which are the magnitudes of **T** and $\mathbf{F}_{ext}$. Thus, Eq. [6] can be solved. We first find the magnitude of the external force. To do this, we rewrite condition 2 in Eq. [6] to isolate T:

$$T = F_{ext} + W_F + F \quad \textbf{(7)}$$

Next we eliminate T from condition 3 in Eq. [6] by inserting Eq. [7]:

$$-\left(F_{ext} + W_F + F\right) l_1 + W_F l_2 + F l_3 = 0 \quad \textbf{(8)}$$

This allows us to determine F_{ext}:

$$F_{ext} = W_F \frac{(l_2 - l_1)}{l_1} + F \frac{(l_3 - l_1)}{l_1} \quad \textbf{(9)}$$

Now the given numerical values are substituted into Eq. [9]. In particular, we note from Table 3.3 that the arm of a standard man has a mass of 0.065 · 70 kg = 4.6 kg and the mass of the lower arm is ½ of this value, i.e., 2.3 kg. We further note that $F = W_M = M \cdot g$:

$$F_{ext} = (2.3\ kg)\left(9.8\ \frac{m}{s^2}\right)\frac{11\ cm}{4\ cm} + (6\ kg)\left(9.8\ \frac{m}{s^2}\right)\frac{36\ cm}{4\ cm} = 590\ N \quad \textbf{(10)}$$

Solution to part (b): F_{ext} from Eq. [10] is substituted into Eq. [7], yielding T:

$$T = F_{ext} + F + W_F$$

$$= (590\ N) + (6\ kg + 2.3\ kg)\left(9.8\ \frac{m}{s^2}\right) \quad \textbf{(11)}$$

$$= 670\ N$$

To get an idea how large this force is, we divide it by 9.8 m/s², yielding almost 70 kg. Thus, the tendon must withstand a force in this problem that is as large as the force you need to lift a 70 kg object!

Problem 5.11

A standard man holds an object of mass m = 2 kg on the palm of the hand with the arm stretched, as shown in Fig. 5.53. Use the torque equilibrium equation to determine the magnitude of the force **F** that is exerted by the biceps muscle, when a = 35 cm, b = 5 cm and the angle $\theta = 80^0$.

(a) Neglect the weight of the lower arm.

(b) is the assumption of a negligible mass of the lower arm in part (a) justified?

Solution to part (a): Read this problem carefully. Note that we are asked to quantify only one parameter, which

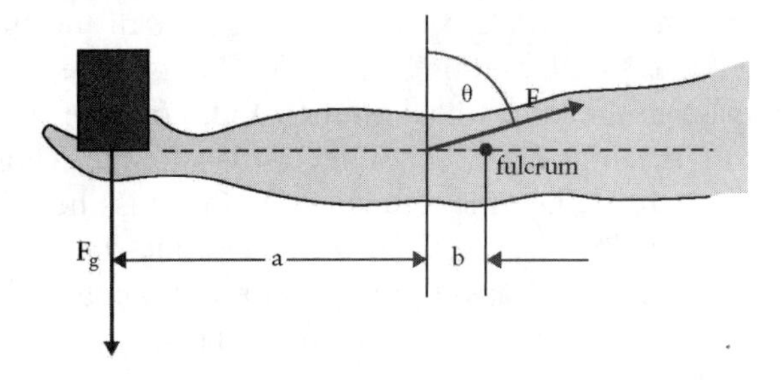

FIGURE 5.53

is the magnitude of the force **F** (its angle with the arm is given). If there is only one unknown variable, only one formula is needed. In the current case, the first two conditions in Eq. [5.27] are not required because the problem text points to the torque equilibrium equation.

The lower arm is the system. It is an extended object with the elbow at a distance $a + b$ to the right of the object held in the hand. Two forces act on the lower arm, the muscle force **F** and the weight of the held object, $\mathbf{F}_g$. The weight of the lower arm itself is neglected.

The balance of torque diagram is shown in Fig. 5.63. The standard x– and y–axes for the balance of torque diagram are included in the figure. Condition 3 in Eq. [5.27] reads:

$$(a + b)\,F_g - b \cdot F\cos\theta = 0 \quad \textbf{(12)}$$

From Eq. [12] F is calculated:

$$F = \frac{(30\ cm)\ (2\ kg)\left(9.8\ \frac{m}{s^2}\right)}{(5\ cm)\ \cos 80^0} = 680\ N \quad \textbf{(13)}$$

Solution to part (b): How do we judge whether a simplification is justified? The best way is to quantify the effect of a neglected term. We know that the mass of the

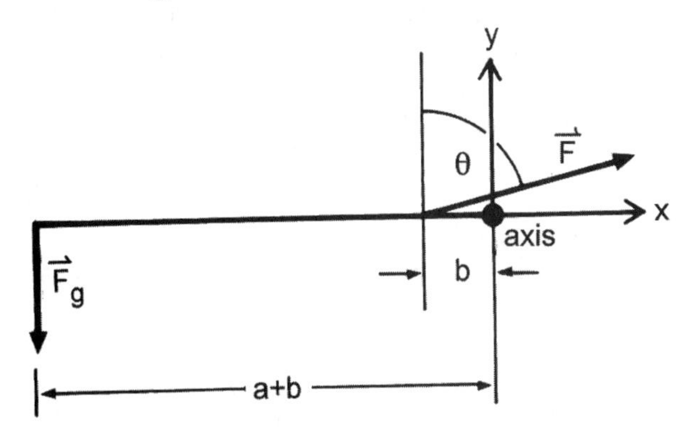

Figure 5.63

lower arm of a standard man is about 50% of the full arm, i.e., 2.3 kg based on Table 3.3. This is more than the mass of the object held on the hand; however, that doesn't matter as the effect of interest is a torque. Thus, we compare τ_m, i.e., the torque due to the mass held in the hand, and $\tau_{neglect}$, i.e., the neglected torque due to the arm. The centre of mass of the lower arm is about at its half–point, a distance ½(a + b) from the fulcrum. We calculate the ratio $\tau_{neglect}/\tau_m$:

$$\frac{\tau_{neglect}}{\tau_m} = \frac{\frac{(a + b)}{2}(2.3\ kg)\ g}{(a + b)\ m\ g} = \frac{2.3\ kg}{2 \cdot 2\ kg} = 0.58 \tag{14}$$

This means that we neglected in part (a) a term that contributes 60% of the term which we used. This problem is therefore an example of an oversimplification.

Problem 5.13

A standard man bends over as shown in Fig. 5.43(a) and lifts an object of mass m = 15 kg while keeping the back parallel with the floor. The muscle that attaches 2/3 of the way up the spine maintains the position of the back. This muscle is called the back muscle or *latissimus dorsi muscle*. The angle between the spine and the force **T** in this muscle is $\theta = 11^0$. Use the balance of torque diagram in Fig. 5.43(b). The weight $\mathbf{W}_1$ includes the object and the arms, the weight $\mathbf{W}_2$ includes the trunk and head of the standard man (use data from Table 3.3).

(a) Find the magnitude of the tension **T** in the back muscle, and

(b) find the x–component of the compressive force **R** in the spine.

<u>Solution to part (a)</u>: We choose the spine as the system. All relevant forces are shown in Fig. 5.43(b), which is the balance of torque diagram; draw the free–body diagram yourself. Then using Eq. [5.27], we find:

condition 1:
$$R_x - T\cos\theta = 0$$

condition 2:
$$R_y + T\sin\theta - W_1 - W_2 = 0 \tag{15}$$

condition 3:
$$L\,W_1 - \frac{2L}{3}T\sin\theta + \frac{L}{2}W_2 = 0$$

in which:

$$W_1 = (m + m_{arms})\ g = (15\ kg + 0.13 \cdot (70\ kg))\left(9.8\ \frac{m}{s^2}\right) = 236\ N \tag{16}$$

and:

$$W_2 = (m_{trunk} + m_{head})\ g = (0.48 + 0.07)(70\ kg)\left(9.8\ \frac{m}{s^2}\right) = 377\ N \tag{17}$$

The magnitude of the tension is obtained from condition 3 in Eq. [15] after division by L:

$$T = \frac{W_1 + \frac{1}{2}W_2}{\frac{2}{3}\sin\theta} = \frac{(236\ N) + \frac{1}{2}(377\ N)}{\frac{2}{3}\sin 11^0} = 3340\ N \tag{18}$$

The tension T is independent of the length L.

<u>Solution to part (b)</u>: The result for T from part (a) is substituted in condition 1 of Eq. [15] to find the component of force **R** that acts along the spine, R_x:

$$R_x = T\cos\theta = (3340\ N)\ \cos 11^0 = 3280\ N \tag{19}$$

Note that condition 2 in Eq. [15] was not needed to solve either part of this problem.

Problem 5.15

A standard man plays on the offence of a football team. Fig. 5.56 shows the quarterback and the centre. Before a play, the centre, the guards and the tackles bend the upper body forward, forming about a $\theta = 45^0$ angle with the horizontal, then remain motionless until the play starts. Fig. 5.57 shows the corresponding balance of torque diagram for the standard man's back. We con-

Figure 5.56

sider the weight of the head H, the arms A, and the trunk T (see Table 3.3), as well as m = 1.2 kg for a typical helmet. Calculate

(a) the magnitude of the tension **T** in the back muscle,

(b) the magnitude of the force $\mathbf{F_B}$ acting on the fifth lumbar vertebra (fulcrum). *Hint*: the figure indicates that the tension **T** forms an 11^0 angle with the spinal column.

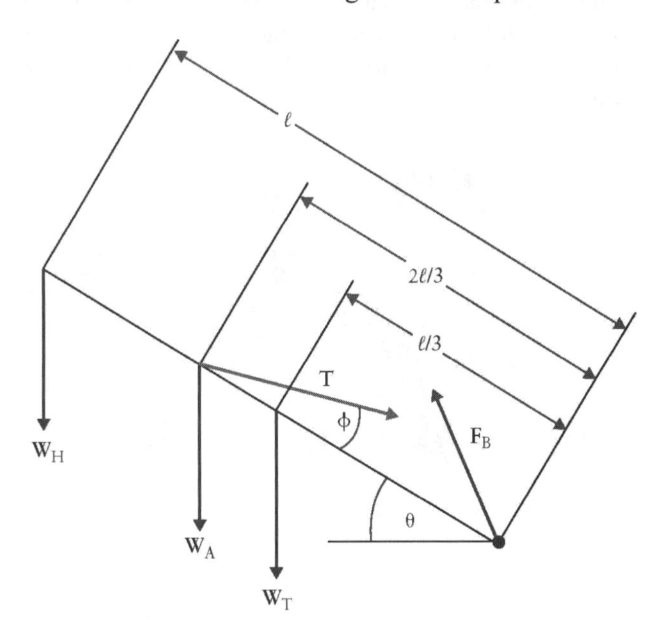

FIGURE 5.57

Solution to part (a): The back of the standard man is the system. Using Fig. 5.57, we identify five forces that act on the system, the three weights $\mathbf{W_T}$, $\mathbf{W_A}$ and $\mathbf{W_H}$, and the two contact forces **T** and $\mathbf{F_B}$. From Fig. 5.57, we further conclude that **T** forms an angle of $\varphi = 34^0$ with the horizontal. The angle of the force $\mathbf{F_B}$ with the horizontal is not known. No numerical value is given for the length l, but the weight of the player is known because he is a standard man.

The free–body diagram is shown in Fig. 5.64. We choose the x– and y–axes as shown. We use the free–body diagram from Fig. 5.64 and the balance of torque diagram from Fig. 5.57 to quantify the problem. Eq. [5.27] is used because the back in the problem is held in rotational equilibrium:

condition 1:

$$-F_{B,x} + T\cos34^0 = 0$$

condition 2:

$$F_{B,y} - T\sin34^0 - W_H - W_A - W_T = 0$$

condition 3:

$$\frac{l}{2}W_T\sin45^0 + l\,W_H\sin45^0 + \frac{2\cdot l}{3}\left(W_A\sin45^0 - T\sin11^0\right) = 0 \qquad \textbf{(20)}$$

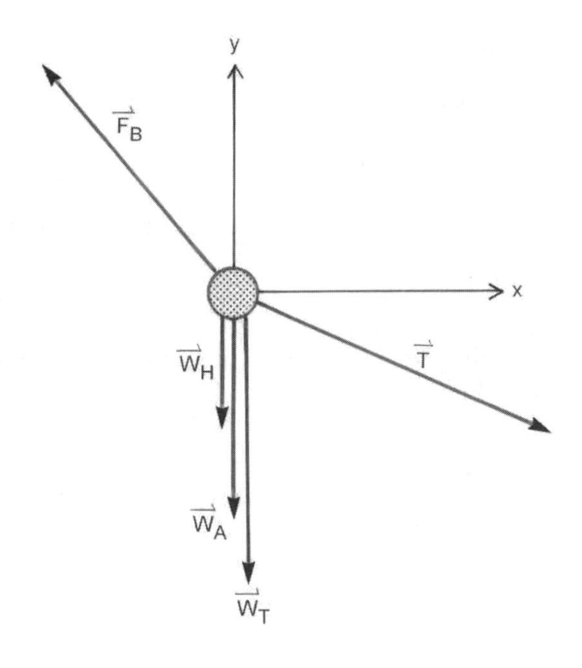

Figure 5.64

Note that conditions 1 and 2 contain two unknown parameters each, condition 1 T and the x–component of $\mathbf{F_B}$, and condition 2 T and the y–component of $\mathbf{F_B}$. Condition 3 contains only T. Thus, the last formula is best suited to solve for part (a) of the problem; we rewrite it to isolate T:

$$\frac{2l}{3}T\sin 11^0 = \frac{l}{2}W_T\sin 45^0 + l\,W_H\sin 45^0 + \frac{2l}{3}W_A\sin 45^0 \quad (21)$$

which yields after division by l:

$$T = \frac{3\sin 45^0}{2\sin 11^0}\left(\frac{W_T}{2} + W_H + \frac{2\,W_A}{3}\right) \quad (22)$$

The factor in front of the bracket on the right side equals 5.56. We substitute:

$$W_T = 0.48\cdot(70\,kg)\left(9.8\,\frac{m}{s^2}\right) = 330\,N$$

$$W_A = 0.13\cdot(70\,kg)\left(9.8\,\frac{m}{s^2}\right) = 90\,N \quad (23)$$

$$W_H = (0.07\cdot 70\,kg + 1.2\,kg)\left(9.8\,\frac{m}{s^2}\right) = 60\,N$$

which yields for Eq. [22]:

$$T = 5.56\left(\frac{330\,N}{2} + 60\,N + \frac{2}{3}\cdot 90\,N\right) = 1580\,N \quad (24)$$

The back muscle must hold 2.3 times the weight of the person when bending down!

Solution to part (b): Substituting the result of part (a) in conditions 1 and 2 of Eq. [20] allows us to determine the two components of the force $\mathbf{F_B}$. The x–component is:

$$F_{B,x} = T\cos 34^0 = (1580\,N)\cos 34^0 = 1310\,N \quad (25)$$

and the y–component reads:

$$F_{B,y} = T\sin 34^0 + W_H + W_A + W_T = (1580\,N)\sin 34^0 + (480\,N) = 1360\,N \quad (26)$$

Using the Pythagorean theorem, Eqs. [25] and [26] yield for the magnitude of force $\mathbf{F_B}$:

$$F_B = \sqrt{F_{B,x}^2 + F_{B,y}^2} = \sqrt{(1310\,N)^2 + (1360\,N)^2} = 1890\,N \quad (27)$$

The tremendous forces acting in the back are the reason why the players in both defensive and offensive lines often rest their weight on one hand while waiting for the snap!

Problem 5.17

Fig. 5.59 shows a standard man wearing a cast supported by a sling, which exerts a vertical force **F** on the lower arm. The distance between the shoulder joint and the elbow is $l_1 = 35$ cm, the mass of the cast is 3 kg. The sling supports the lower arm at the centre of mass $l_2 = 15$ cm from the elbow. Use one–half of the mass of the arm without cast for the mass of the lower arm, and $\theta = 70^0$ for the angle of the arm at the elbow. Calculate the magnitude of force **F**. *Hint*: Other forces act on the arm, which we assume to act along a line through the shoulder joint. Thus, equating **F** with the weights of the upper and lower arm does not yield the correct result. Instead, the problem is solved with the balance of torque equation, in which the unknown forces at the shoulder joint are not included.

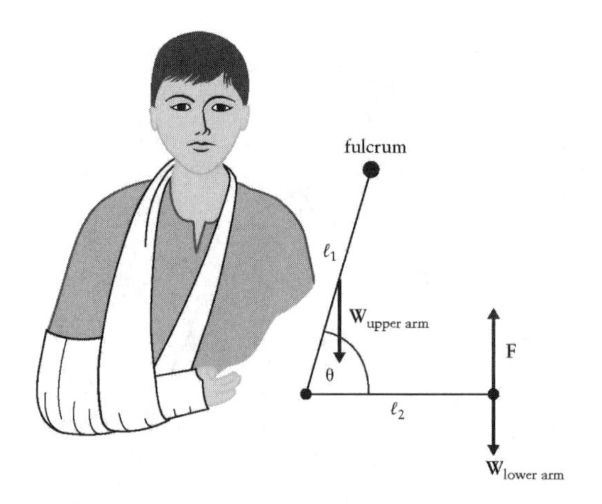

FIGURE 5.59

Solution: The shoulder joint is the fulcrum; some of the forces acting on the arm are exerted below the elbow, i.e., below the point at which the arm is bent. Therefore, this problem is an application of the concepts introduced in Example 5.4.

The problem text states that we need to use the torque equilibrium equation. Note that we cannot draw

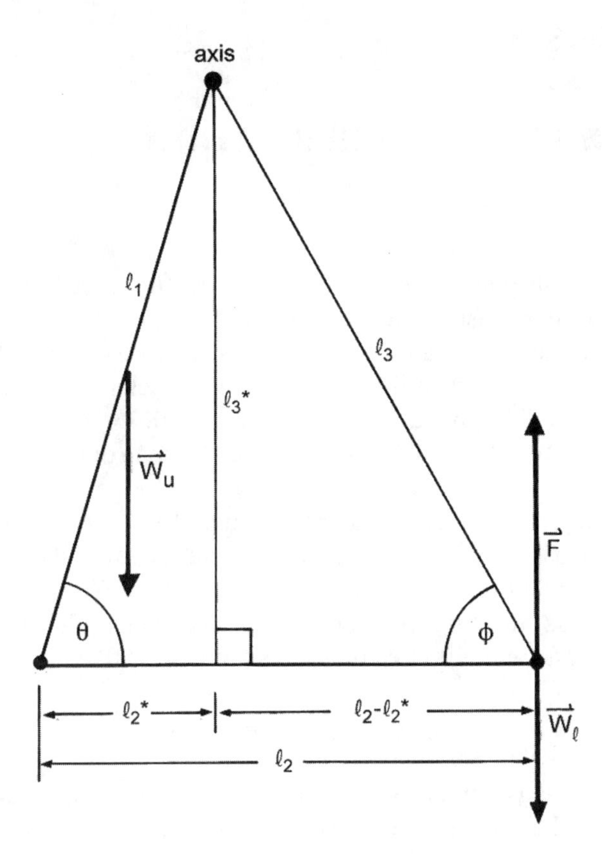

Figure 5.65

a free–body diagram because not all the forces acting on the arm are identified. Using the concepts discussed in Example 5.4 we draw a second balance of torque diagram in Fig. 5.65 which defines several additional distances and angles. In addition to the three forces and two lengths shown in Fig. 5.59, a line of length l_3 is added from the shoulder to the centre of mass of the lower arm, and a new angle φ is defined. For the purpose of geometrical construction, the distance l_3* is introduced which is the length of the line that is perpendicular to the lower arm and runs through the fulcrum. With these additional variables introduced, condition 3 in Eq. [5.27] is written:

$$W_u \sin(90^0 - \theta)\frac{l_1}{2} + F\sin(90^0 - \varphi)\, l_3 - W_l \sin(90^0 - \varphi)\, l_3 = 0 \qquad \textbf{(28)}$$

Eq. [28] contains not only F as an unknown variable, but also l_3 and φ. Thus, before we can proceed with solving Eq. [28], these two variables must be determined. l_3 is found with the Pythagorean theorem once the lengths l_3* and $l_2 - l_2$* are calculated. Trigonometry shows that the following three relations hold:

(i):

$$l_3^* = l_1 \sin\theta = (35\ cm)\sin 70^0 = 33\ cm \qquad \textbf{(29)}$$

(ii):

$$l_2^* = l_1 \cos\theta = (35\ cm)\cos 70^0 = 12\ cm \qquad \textbf{(30)}$$

and (iii):

$$l_2 - l_2^* = 8\ cm \qquad \textbf{(31)}$$

and thus:

$$l_3 = \sqrt{(l_3^*)^2 + (l_2 - l_2^*)^2} = \sqrt{(33\ cm)^2 + (8\ cm)^2} = 34\ cm \qquad \textbf{(32)}$$

Using Fig. 5.65, we find φ:

$$\tan\varphi = \frac{l_3^*}{l_2 - l_2^*} = \frac{33\ cm}{8\ cm} = 4.1 \quad \Rightarrow \quad \varphi = 76^0 \qquad \textbf{(33)}$$

Eq. [28] is solved with Eqs. [30] to [33] to find F. We use for the two weights each (4 kg) (9.8 m/s²) = 39 N:

$$F = 10.6\ N \qquad \textbf{(34)}$$

This small force is intentional as a large force would unnecessarily strain the patient's neck.

CHAPTER SIX

Bioenergetics: Energy and its Conservation

MULTIPLE CHOICE AND CONCEPTUAL QUESTIONS

Question 6.1
(a) Which is the standard unit for energy? (A) N, (B) J/s, (C) N · m, (D) Pa, (E) N/J.
(b) Which is the standard unit for pressure? (A) N/s, (B) kg / (m · s^2), (C) J/s, (D) N · m, (E) Pa · m^3.
(c) Work is measured in the same unit as (A) force, (B) energy, (C) pressure, (D) momentum, (E) power.

Answer to part (a): (C)
Answer to part (b): (B)
Answer to part (c): (B)

Question 6.3
A force of 5 N causes the displacement of an object by 3 m in the direction of the force. What work did the object do/has been done on the object? (A) W = + 1.5 J, (B) W = – 1.5 J, (C) W = + 15.0 J, (D) W = – 15.0 J, (E) none of the above.

Answer: (C)

Question 6.5
A person pushes an object off a seat and then tries to lift it with an external force **F** upward into an overhead bin. However, the person is too weak and the object drops to the floor under its own weight **W**, pulling the person along as distance Δy. Which of the following statements is correct? (A) Because the person did not succeed, no work is done in this experiment. (B) The person has done work on the object; the absolute value of the work is W = |**F**| · Δy. (C) The person has done work on the object; the absolute value of the work is W = |**W**| ·Δy. (D) The object has done work on the person; the absolute value of the work is W = |**W**| · Δy. (E) The object has done work on the person; the absolute value of the work is W = |**F**| · Δy.

Answer: (E)

Question 6.7
Air, initially at 100 kPa, is sealed in a container by a mobile piston of cross–sectional area 10.0 cm^2. Now we push the piston with an additional force of F = 100 N to compress the air. What is the final pressure p of the sealed air when the piston reaches mechanical equilibrium? (A) p = 2 × 10^4 Pa, (B) p = 1.0001 × 10^5 Pa, (C) p = 2 × 10^5 Pa, (D) p = 1 × 10^6 Pa, (E) none of the above.

Answer: (C). Remember to change the cross–sectional area into standard units. The pressure caused by the additional force is p = F/A = (100 N) / (0.001 m^2) = 100 kPa. Thus, the final pressure in the container is the initial pressure of 100 kPa plus the additional 100 kPa.

Question 6.9
A gas is compressed from 5 L to 1 L with its pressure held constant at 4000 Pa. To achieve this compression, the following work (absolute value only) is needed: (A) W = 20 kJ, (B) W = 16 kJ, (C) W = 20 J, (D) W = 16 J, (E) none of the above.

Answer: (D). The volume decreases by ΔV = 4 L = 0.004 m^3. Thus, W = – p · ΔV = – (4000 Pa) (– 0.004 m^3) = 16 J.

Question 6.11
How does the kinetic energy of an object change when its speed is reduced to 50% of its initial value? (A) The kinetic energy remains unchanged. (B) The kinetic energy becomes 50% of the initial value. (C) The kinetic energy becomes 25% of the initial value. (D) The kinetic energy doubles. (E) The kinetic energy increases fourfold.

Answer: (C)

Question 6.13
How does the potential energy of an object change when its speed is doubled? (A) The potential energy is halved. (B) The potential energy becomes 1/4 of the initial value. (C) The potential energy doubles. (D) The potential energy increase by a factor of four. (E) The information given is insufficient to choose one of the above answers.

Answer: (E). Increasing the speed will increase the object's kinetic energy, but we do not know that the increase in kinetic energy is related to a decrease in potential energy. Even if the sum of kinetic and potential energy of the system were conserved answer (E) remains correct because we can only quantify changes in potential energy but the absolute value is arbitrary.

Question 6.15
In a mechanical experiment with an isolated object, only the values of E_{kin} and E_{pot} can vary. If the object accelerates from 5 m/s to 10 m/s, its potential energy has: (A) not changed, (B) decreased by a factor of 2, (C) decreased by a factor of 4, (D) decreased by a factor we cannot determine from the problem as stated, (E) increased.

Answer: (D). Like in Question 6.13, the absolute value of the potential energy depends on the choice of origin along the y–axis. It can therefore have any possible value initially.

Question 6.17
If the speed of a particle is doubled, what happens to its kinetic energy?

Answer: Since $E_{kin} \propto v^2$, the kinetic energy must increase by a factor of 4.

Question 6.19
In pole vault several forms of energy play a role: kinetic energy of the runner, elastic potential energy of the pole (discussed in Chapter 16), the gravitational potential energy, and the internal energy of the vaulter, which is associated with muscles, tendons, and ligaments. For this question, however, we simplify the discussion by neglecting all but the kinetic and gravitational potential energy. We want to estimate what the highest possible pole vault is if the centre of mass of the athlete is 1.1 m above the ground and the maximum speed of approach is 11 m/s. Use the conservation of energy to estimate this height, choose the closest value. (A) 6.2 m, (B) 7.3 m, (C) 11.0 m, (D) 14.6 m.

Answer: (B). Using conservation of energy, we find the maximum increase in height:

$$h = \frac{v^2}{2g} = \frac{\left(11\ \frac{m}{s}\right)^2}{2\left(9.8\ \frac{m}{s^2}\right)} = 6.2\ m \qquad \textbf{(1)}$$

This is the increase in the height of the centre of mass. This means that the bar can be at a maximum height of 6.2 m + 1.1 m = 7.3 m.

Question 6.21
Which of the following formulas describes the reading of a temperature T (in degrees Celsius) when measured with an expanding mercury column as proposed by Celsius? In these formulas, h is the height of the mercury column and H is the height of the mercury column at a chosen reference temperature. α and β are positive constants, with α the reference temperature at which the column height is $h = H$.

$$\begin{aligned}
&(A)\quad T = \alpha - \beta\left(\frac{h-H}{H}\right)\\
&(B)\quad T = \alpha + \beta\left(\frac{h-H}{H}\right)\\
&(C)\quad T = \alpha - \beta\left(\frac{h+H}{H}\right)\\
&(D)\quad T = \alpha + \beta\left(\frac{h+H}{H}\right)\\
&(E)\quad T = \alpha - \beta\left(\frac{h+H}{H}\right)^2
\end{aligned} \qquad \textbf{(2)}$$

Answer: (B). Choices (C), (D) and (E) do not result in $T = \alpha$ at $h = H$. Choice (A) implies $T \propto -h$, i.e., the mercury column would shrink for rising temperatures.

Question 6.23
In the next chapter a new temperature scale (Kelvin scale with unit K) is introduced, with $T(K) = T(^0C) + 273.15^0$. In an astronomy class, the temperature at the core of a star is reported as 1.5×10^7 degrees. Does it make sense to ask whether the temperature scale used is in units ^{0}C or K?

Answer: In the case of such an astronomically large temperature, the difference is less than 0.002%. Chances are good that such a small difference would be insignificant. But, check out the percentage error you would get when reporting the temperature that water freezes if you need to convince yourself that it is important to know which unit is being used.

Question 6.25
Which of the following things cannot happen in a closed system? (A) Heat is transferred to the environment. (B) Work is done on the system by an object in the environment. (C) Matter flows into the system. (D) The temperature of the system increases. (E) The internal energy of the system remains unchanged.

Answer: (C)

Question 6.27
Early Europeans arriving in North America stored fruit and vegetables in underground cellars. In winter they also included a large open barrel with water. Why did they do this?

Answer: In cold winters the heat stored in the barrel of liquid water helped keep the temperature of the storage room above freezing.

Question 6.29
The air temperature above costal areas is significantly affected by the large specific heat of water. Estimate the amount of air for which the temperature can rise 1 degree if 1 m³ of water cools by 1 degree. The specific heat of air is 1.0 kJ/(kg · °C); for its density use 1.3 kg/m³.

Answer: The heat goes from the water to the air, thus:

$$Q = c_w \rho_w V_w \Delta T = c_{air} \rho_{air} V_{air} \Delta T \qquad \textbf{(3)}$$

Solving for the volume of air gives us:

$$V_{air} = \frac{c_w \rho_w V_w}{c_{air} \rho_{air}} = \frac{(4.2 \frac{kJ}{kg})(1000 \frac{kg}{m^3})(1\ m^3)}{(1 \frac{kJ}{kg})(1.3 \frac{kg}{m^3})} \qquad \textbf{(4)}$$

Which yields $V_{air} = 3200\ m^3$.

Question 6.31
What is wrong with stating that, of any two objects, the one with the higher temperature contains more thermal energy?

Answer: Heat capacities and masses of the objects have been neglected.

ANALYTICAL PROBLEMS

Problem 6.1
Fig. 6.27(b) shows in double–logarithmic representation the nerve impulse rate P as a function of the speed of an approaching object for a Meissner's corpuscle. Using the power law relation $P = a \cdot v^b$, determine the constants a and b.

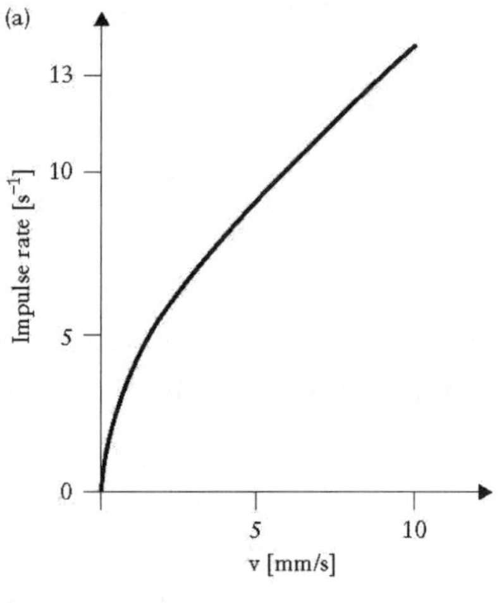

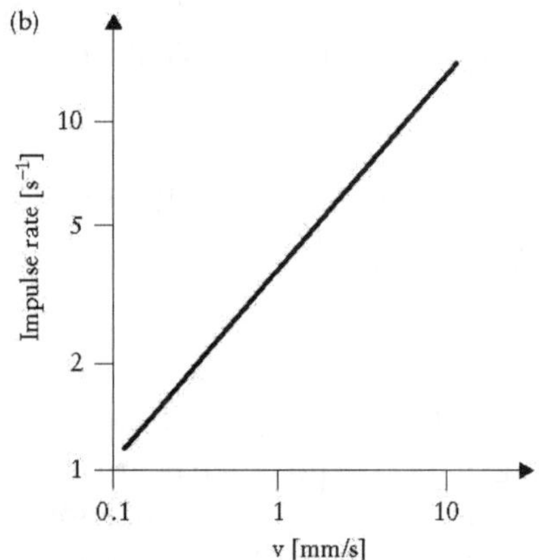

FIGURE 6.27

Solution: Fig. 6.27(b) is a double–logarithmic plot. As outlined in the Math Review *Graph Analysis Methods* on p. 9 – 12 in Chapter 1, we start with supplementing both axes with linear axes that show values for ln$P$ and ln$v$. After adding these additional axes the power law is re-written in logarithmic form:

$$P = a \cdot v^b \Rightarrow \ln P = \ln a + b \ln v \qquad \textbf{(5)}$$

Eq. [5] is analysed to obtain the constants a and b. From Fig. 6.27(b) we read two data pairs for ln$P$ and ln$v$:

Table 6.7

Data set (Units)	$\ln P$ P in s^{-1}	$\ln v$ v in mm/s
#1	0.693	−1.099
#2	2.303	+1.714

The two data sets in Table 6.7 are chosen by reading the speed values at impulse rates of $P = 2\ s^{-1}$ and $P = 10\ s^{-1}$. Table 6.7 allows us to write two linear formulas based on Eq. [5]:

$$\begin{aligned} (I) \quad & 0.693 = b \cdot (-1.099) + \ln a \\ (II) \quad & 2.303 = b \cdot (+1.714) + \ln a \\ \hline (II) - (I) \quad & 1.61 = b \cdot (1.714 + 1.099) \end{aligned} \tag{6}$$

Solving Eq. [6] for b yields $b = 0.57$. A value for $\ln a$ is obtained by substituting this result in either of the two formulas in Eq. [6]. This yields $\ln a = 1.326$, which means that $a = 3.8$.

Note that we did not identify a physical concept (or law) which justifies the power law in Eq. [5], but introduced this relation empirically. This is often done in the sciences and allows you to develop an idea of the physical processes underlying the data without having established firm physical laws. There are, of course, several differences between an empirical relation and a physical law. For example, the coefficients a and b have been calculated from experimental data and carry an uncertainty; their true values can only be identified through a physical reasoning for the law relating the impulse rate and the speed of the approaching object. This distinction is important when we ask for the units of the two coefficients. As long as we use an empirical relation, the units of a and b are meaningless and should not be considered further; if we had a proposed physical law, then the units of the coefficients would have to be consistent with the model.

Problem 6.3

In the mid–Cretaceous (110 to 100 million years ago), dinosaurs lived near the poles, e.g., at 80^0N for fossils in North Alaska and the Yukon, and at 80^0S for fossils near Melbourne, Australia. The polar regions of the Cretaceous were densely forested with only occasional light freezes in the winter, but non–hibernating ectotherms cannot tolerate prolonged periods without sunlight. Indeed, the most northern fossil find of a large ectotherm is a giant crocodile (*Phobosuchids*) at 55^0N. The energy cost for long–distance migration across land is given in unit J/m as a function of the animal's body mass m in kg:

$$E_{migration} = 14 \cdot m^{3/4} \tag{7}$$

(a) Using Eq. [6.8]:

$$\begin{aligned} \textit{ectotherm:} \quad & M = 20 \cdot m^{3/4} \\ \textit{endotherm:} \quad & M = 450 \cdot m^{3/4} \end{aligned} \tag{8}$$

and Eq. [7] evaluate the hypothesis that ectothermic southern polar dinosaurs migrated annually between 80^0S and 55^0S latitudes for (I) Leaellynasaura, which was a 10-kg herbivore, (II) Dwarf Allosaur, which was a 500-kg carnivore, and (III) Muttaburrasaurus, which was a 4-tonne herbivore. Note in this context that caribou in the Canadian North migrate 4000 km annually.
(b) Using the energy consumption for migration in Eq. [7] and the potential energy, compare the benefits of living in plains versus mountainous terrain for small and large endotherms.

Solution to part (a): The problem is solved in the following steps: first, the length of the migration is calculated. Eq. [7] is then used to determine the metabolic requirement for such a migration. Eq. [8] is used to calculate the annual energy production for the same animal. The migration can only occur if its energy requirement is less than the energy production during the same time period.

The distance associated with an annual migration between 80^0S and 55^0S is determined from the radius of Earth. With $R_{Earth} = 6370$ km, Earth's circumference is $C = 2 \cdot \pi \cdot R_{Earth} = 40000$ km. This circumference corresponds to 360^0; the assumed migration D covered 50^0 (back and forth), which equals:

$$\begin{aligned} D &= (4 \times 10^7\ m)\ \frac{50^0}{360^0} \\ &= 5.5 \times 10^6\ m = 5500\ km \end{aligned} \tag{9}$$

We multiply this distance with the energy consumption per metre as given in Eq. [7]:

$$E_{required} = E_{migration} \cdot D \tag{10}$$

which yields:

$$E_{required} = \left(14 \frac{J}{m}\right) m^{3/4} (5.5 \times 10^6 m) = (7.7 \times 10^7 J)(m (kg))^{3/4} \quad (11)$$

The annual metabolic rate of an ectotherm of the same mass is calculated from Eq. [8]:

$$\mathsf{M} = 7.3 \times 10^6 (m (kg))^{3/4} \quad (12)$$

From Eqs. [11] and [12] we determine the ratio of energy required to energy available per year:

$$\frac{E_{required}}{\mathsf{M}} = \frac{7.7 \times 10^7 \cdot m^{3/4}}{7.3 \times 10^6 \cdot m^{3/4}} = 10.5 \quad (13)$$

Note that this result is independent of the mass of the individual species we study. Thus, ectothermic dinosaurs of any size could not have migrated as assumed because the migration would require more than 10 times their annual metabolic rate. Modern endothermic predators may use up to 25% of their metabolic rate walking in search of food.

Solution to part (b): Eq. [7] quantifies the energy consumption when walking long distances on flat ground. It states that the energy needed per metre travelled is proportional to $m^{3/4}$. In mountainous terrain, a vertical travel component has to be included. The change in potential energy per metre height difference is $E_{pot}/h \propto m \cdot g$, i.e., it is directly proportional to the mass of the animal. Thus, the ratio of energy required for horizontal versus vertical travel for the same animal is:

$$\frac{E_{horizontal}}{E_{vertical}} \propto \frac{m^{3/4}}{m} = m^{-1/4} \quad (14)$$

i.e., lighter animals are favoured when a significant component of vertical movement is involved.

Problem 6.5

We considered in Concept Question 6.1 a model in which an animal's metabolic requirements are determined by the loss of heat through the skin. Why does this lead to $b = 2/3$ as the exponent in $\mathsf{M} \propto m^b$ with m the mass of the animal and M the metabolic rate?

Solution: The surface of an animal A is proportional to r^2 if r is its size (its length in unit metre). The mass of the animal is proportional to its volume, $m \propto V$, and in turn, the volume of the animal V is proportional to r^3. The energy loss ΔE through the surface is proportional to the surface A, i.e., $\Delta E \propto A$ and therefore $\Delta E \propto r^2$. We want to express this last relation as a function of volume, not radius. Therefore, we use $r \propto V^{1/3}$. Substituting this leads to $\mathsf{M} \propto \Delta E \propto r^2 \propto (V^{1/3})^2 = V^{2/3} \propto m^{2/3}$.

Problem 6.7

(a) The highest head–first dive is preformed by professional divers near Acapulco, Mexico, from a height of 35 metres (see Fig. 6.40). With what speed does a diver (standard man) enter the water if leaving the platform from rest?

(b) Draw the speed, the kinetic and the potential energy as a function of height for the diver. *Hint*: Neglect air resistance.

Figure 6.40

Solution to part (a): We neglect air resistance; if you want to evaluate that assumption further, read Example 4.6. The diver is not in mechanical equilibrium as a continuous acceleration occurs. We use the conservation of energy with the initial state of the system when the diver leaves the platform and the final state when the diver reaches the water surface. The initial state is chosen because we know all required parameters at that point: the diver starts from rest and is 35 m above the water surface. The final state is chosen at the water surface because the unknown parameter, the final speed v_{final}, applies at that point. We write:

$$E_{kin,\ initial} + E_{pot,\ initial} = E_{kin,\ final} + E_{pot,\ final} \quad \textbf{(15)}$$

The individual terms for the diver are:

$$\begin{aligned} E_{kin,\ initial} &= 0 \\ E_{pot,\ initial} &= m \cdot g \cdot h \\ E_{kin,\ final} &= \frac{1}{2} m \cdot v_{final}^2 \\ E_{pot,\ final} &= 0 \end{aligned} \quad \textbf{(16)}$$

The initial kinetic energy is zero because the diver starts from rest. The final potential energy is zero since we choose $y = 0$ at the surface of the water. Therefore, the initial height is $y = h = 35$ m. We substitute the four formulas from Eq. [16] in Eq. [15]:

$$m \cdot g \cdot h = \frac{1}{2} m \cdot v_{final}^2 \quad \textbf{(17)}$$

which yields:

$$v_{final} = \sqrt{2 \cdot g \cdot h} \quad \textbf{(18)}$$

Note that Eq. [18] does not contain the mass of the system, i.e., it applies to all divers, whether they confirm to standard man data or not. Substituting the values given in the problem text we find:

$$\begin{aligned} v_{final} &= \sqrt{2\left(9.8\,\frac{m}{s^2}\right) 35\ m} \\ &= 26\,\frac{m}{s} = 94\,\frac{km}{h} \end{aligned} \quad \textbf{(19)}$$

This result indicates why these dives are only done by professional athletes: you don't want to hit the water surface with that speed unless your body is perfectly aligned for the dive. Compounding the risk in Mexico is the fact that the water at the bottom of the cliff is rather shallow and the divers have to ensure that a wave has rolled in when they enter the water.

Solution to part (b): Fig. 6.44 shows a sketch of the diver and the corresponding energy plots (centre) and plots of speed versus height (left). The total mechanical energy is constant and the potential energy increases linearly with height. Due to energy conservation, the kinetic energy decreases linearly with height, $E_{kin} = ½\ m \cdot v^2 \propto$ const $- y$. The square–root of this relation describes the speed as a function of height, i.e., $v \propto (\text{const} - y)^{1/2}$. Note that the graph of velocity versus time for an object in free–fall is a parabola, whereas this curve is a square root. This is because we are plotting versus distance in this problem.

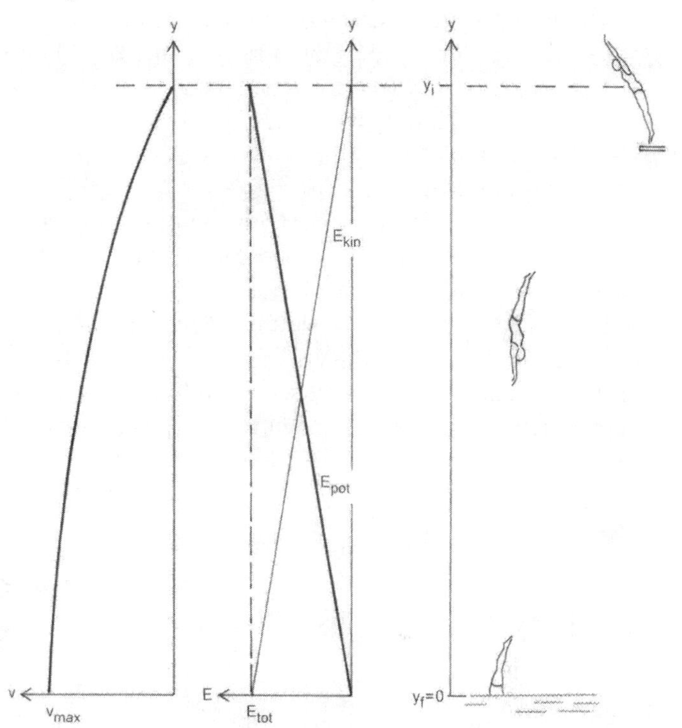

Figure 6.44

Problem 6.9

Three objects with masses $m_1 = 5.0$ kg, $m_2 = 10.0$ kg, and $m_3 = 15.0$ kg are attached by massless strings over two frictionless pulleys, as shown in Fig. 6.42. The horizontal surface is frictionless and the system is released from rest. Using energy concepts, find the speed of m_3 after it has moved down 0.4 m.

Solution: We apply the conservation of energy. The initial instant is chosen at the release of the objects from rest, and the final instant when mass m_3 has moved 0.4 m downward. For each of the three objects, we identify all

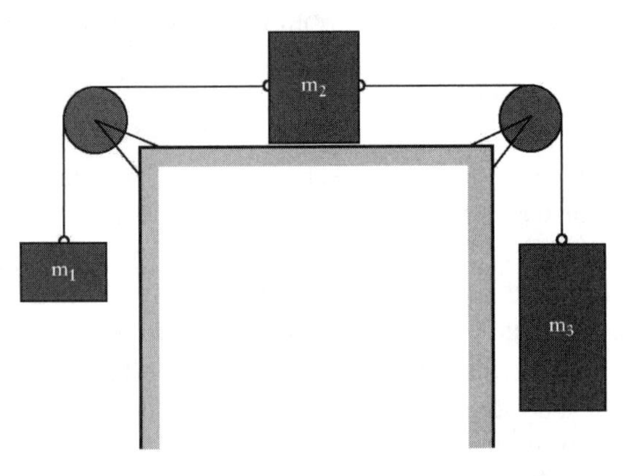

FIGURE 6.42

the terms in Eq. [15] separately:

- The initial kinetic energy of all three objects is zero since they are connected by taut strings. When one of the objects is at rest, the other two must be at rest, too.
- The initial potential energy we choose to be zero by defining the initial position of each object as the respective origin along the vertical axis.
- The final kinetic energy contains the unknown variable. Thus, we leave that term unchanged in Eq. [15] for now.
- The final potential energy has three contributions, one due to each of the three objects. Since the object of mass m_2 moves across a horizontal surface only, its potential energy does not change; the other two objects, however, contribute:

$$E_{pot,final} = m_1 g y_{1,final} + m_3 g y_{3,final}$$
$$= \left(5\,kg\,(0.4\,m) + 15\,kg\,(-0.4\,m)\right)\left(9.8\,\frac{m}{s^2}\right) \quad \textbf{(20)}$$

where a downward displacement of $\Delta y_3 = -0.4$ m for the object of mass m_3 means an upward displacement of the object of mass m_1 because the connecting strings are taut: $\Delta y_1 = +0.4$ m. This yields:

$$E_{pot,\,final} = -39.2\,J \quad \textbf{(21)}$$

Next we enter all energy contributions at the initial and final states of the system into Eq. [15]:

$$0 = -39.2\,J + \frac{1}{2}\left(m_1 + m_2 + m_3\right) v_{final}^2 \quad \textbf{(22)}$$

There is only one speed, v_{final}, associated with the total final kinetic energy because all three objects move at the same speed at all times. The final speed is:

$$v_{final} = \sqrt{\frac{2\,(39.2\,J)}{(5 + 10 + 15)\,kg}} = 1.62\,\frac{m}{s} \quad \textbf{(23)}$$

Problem 6.11

If a person lifts a bucket of mass 20 kg from a well and does 6.0 kJ work, how deep is the well? Assume that the speed of the bucket is constant as it is lifted.

Solution: The work is done in changing the potential energy of the bucket. To find the depth of the well use $h = E_{pot} / (m \cdot g) = 6000\ J / (20\ kg \cdot 9.8\ m/s^2) = 30.6$ m.

Problem 6.13

A standard man executes a pole vault. The approach is at 10 m/s and the speed moving above the bar is 1.0 m/s. Neglecting air resistance and energy absorbed by the pole, determine the maximum height of the bar.

Solution: The change in kinetic energy of the vaulter is given by $\Delta E_{kin} = ½\, m\,(v^2_{final} - v^2_{initial}) < 0$. Equate $-E_{kin}$ to the potential energy at the maximum height to find the relation $½\, m\,(v^2_{initial} - v^2_{final}) = m \cdot g \cdot h$. Solve to find h. Notice that the answer will be independent of the mass of the vaulter:

$$h = \frac{\left(v_{initial}^2 - v_{final}^2\right)}{2 \cdot g} = 5.1\,m \quad \textbf{(24)}$$

Problem 6.15

A gas expands from a volume of 1.0 L to 5.0 L, as shown in the p–V diagram of Fig. 6.43. How much work

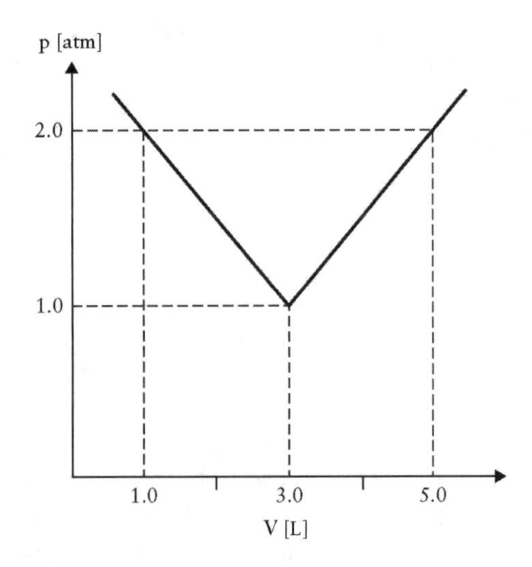

FIGURE 6.43

does the gas perform on the piston?

Solution: The definition of work is based on the area under the curve representing a process in a p–V diagram. If the process occurs at constant pressure, $p = \text{const} = p_0$, the work is calculated as:

$$W = -p_0 \left(V_{final} - V_{initial}\right) \quad \textbf{(25)}$$

For any other process, either graphic or numerical methods are needed to obtain the area. Fig. 6.43 represents an intermediate case: the pressure varies along straight line segments. In this case, a geometrical method can be applied to obtain the work.

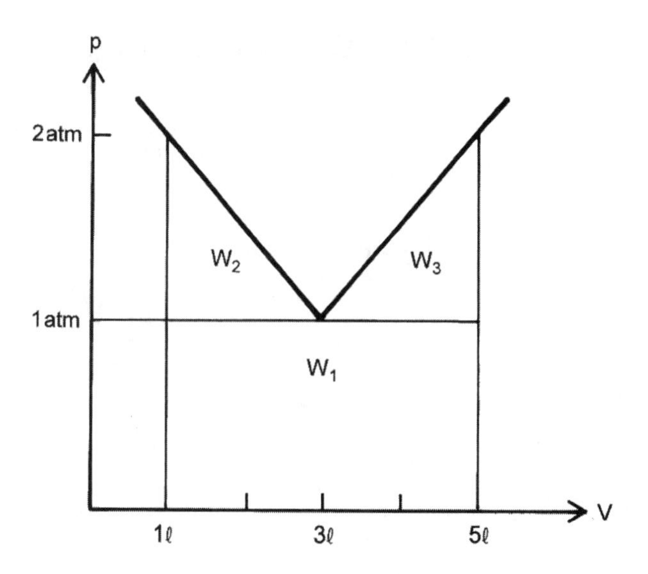

Figure 6.45

The area under the curve in Fig. 6.43 is divided into three simple–shaped areas, as shown in Fig. 6.45. We calculate the contribution to the total work of each of these three areas; the total work is then the sum of the three components. Using geometric identities for right triangles and rectangles, we find for the three areas:

$$\begin{aligned} A_1 &= (1\,atm)\,(5\,L - 1\,L) \\ &= 4\,atm \cdot L \\ A_2 &= \frac{1}{2}(2\,atm - 1\,atm)\,(3\,L - 1\,L) \\ &= 1\,atm \cdot L \\ A_3 &= \frac{1}{2}(2\,atm - 1\,atm)\,(5\,L - 3\,L) \\ &= 1\,atm \cdot L \end{aligned} \quad \textbf{(26)}$$

The terms in Eq. [26] were intentionally not labelled W_1, W_2 and W_3 because we must still apply the sign convention. The work is negative because the process in Fig. 6.43 is an expansion for which work is done by the gas on the piston. Thus, the total work is:

$$\begin{aligned} W &= -\left(A_1 + A_2 + A_3\right) \\ &= -6\,atm \cdot L = -608\,J \end{aligned} \quad \textbf{(27)}$$

The unit $atm \cdot L$ in Eq. [27] is not an SI energy unit and thus, has been converted with 1 L = 1×10^{-3} m^3 and 1 atm = 1.013×10^5 Pa.

Note that Fig. 6.43 does not represent a realistic process. However, it is often possible to approximate a real process with a few straight line segments. Thus, the method above is frequently used to obtain a first approximation of the work in a real situation.

Problem 6.17

Table 6.5 shows the metabolic rate for given activities of the adult human body and Table 6.6 gives the energy content of the three most important components of food.
(a) How much energy is expended by a standard man who walks for one hour every morning?
(b) If the body of the person consumes body fat reserves to produce this energy, how much mass will be lost per day?

Table 6.5

Activity	Metabolic rate ($cal/(s \cdot kg)$)
Sleeping	0.263
Sitting	0.358
Standing	0.621
Walking	1.0
Biking	1.81
Swimming	2.63
Running	4.3

Table 6.6

Food	Energy content (cal/g)
Carbohydrate	4100
Protein	4200
Fat	9300

Solution to part (a): From Table 6.5, we find the metabolic rate for walking to be M = 1.0 cal/(s · kg). Note that the non–standard unit cal is often used in the context of energy content in food or other organic compounds. You also find the unit Cal (with a capital letter C) which converts as 1 Cal = 1 kcal = 1×10^3 cal. It is advisable to convert such non–standard units as early as possible. Here the result is M = 1.0 cal/(s · kg) = 4.184 J/(s · kg).

From this value of the metabolic rate M we determine the total energy when a standard man of 70 kg walks for 1 h = 3600 s. Note that the non–standard unit h (hour) is used. The energy consumption ΔE_{walk} is:

$$\Delta E_{walk} = \left(4.184 \frac{J}{s\ kg}\right) (70\ kg)\ (3600\ s) = 1.05 \times 10^6\ J = 1050\ kJ \quad \textbf{(28)}$$

Solution to part (b): Using the result in part (a) and Table 6.6, the fat consumption during the walking exercise is calculated. The table shows that 1 g = 0.001 kg fat provides 9300 cal = 39 kJ when converted into energy. Thus, the total mass of fat consumed is:

$$m_{fat\ loss} = \frac{1050\ kJ}{39\ kJ/g} \left(1 \times 10^{-3} \frac{kg}{g}\right) = 0.027\ kg \quad \textbf{(29)}$$

i.e., only 27 g of fat is lost.

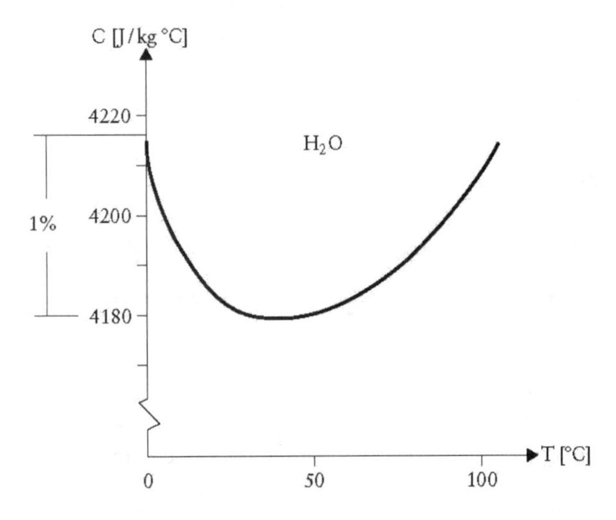

FIGURE 6.35

Problem 6.19

Assume that Joule's brewery horses each did $P = 750$ J/s of work per second (this corresponds roughly to the definition of horsepower). If Joule had 4 horses moving in a circle for 1 hour to operate a stirrer in a well–isolated container filled with 1 m³ water at an initial temperature of 25°C, to what final value did the water temperature rise? Use Fig. 6.35 for the specific heat of water at 25°C.

Solution: 4 horses perform work at a rate 4 · 750 J/s = 3000 J/s. If the total work is stored in the water in the form of thermal energy as implied in the problem text, the equivalence of work and heat applies in the form:

$$W = P \cdot \Delta t = m_{H_2O}\, c_{H_2O}\, \Delta T \quad \textbf{(30)}$$

The temperature difference is obtained from Eq. [30]:

$$\Delta T = \frac{P \cdot \Delta t}{\rho_{H_2O}\, V_{H_2O}\, c_{H_2O}} = \frac{\left(3000 \frac{J}{s}\right)(3600\ s)}{\left(1 \times 10^3 \frac{kg}{m^3}\right)(1.0\ m^3)\left(4.185 \frac{kJ}{°C\ kg}\right)} = 2.6°C \quad \textbf{(31)}$$

in which ρ_{H2O} is the water density with ρ_{H2O} = 1.0 g/cm³ and the heat capacity of water at 25°C from Fig. 6.35 is c_{H2O} = 4185 J/(°C · kg), which, after unit conversion, is equal to c_{H2O} = 4185 J/(°C · kg). The temperature change in Eq. [31] is measurable but small, increasing from 25°C to 27.6°C.

Problem 6.21

The energy extracted from burning sugar is used in the mitochondria to synthesize ATP from ADP. Consider glucose, which releases 675 kcal/mol during cellular respiration (formation of CO_2). What fraction of this energy is used in the ATP formation if 38 molecules of ATP are formed for each molecule of glucose? Why is this value not 100%?

Solution: The reverse amount of energy listed in Eq. [6.2] is required in the phosphorylation of ADP to ATP: – 29 kJ/mol. Thus, the formation of 38 mol of ATP requires 1100 kJ. The amount released per mol in the cellular respiration of glucose is 675 kcal = 2825 kJ. This means that 1100 kJ/2825 kJ = 39% of the energy stored in the glucose molecule can be used to synthesize ATP from ADP.

The mitochondria work like small power plants,

i.e., they operate as an open system applying a cyclic process. It must be a cyclic process because otherwise they would do their deed once and would become obsolete in no time. Read the discussion about the second law of thermodynamics in Chapter 8 to learn more about the efficiency of cyclic processes and their inadvertent loss of thermal energy during operation.

Problem 6.23

Water at the top of Niagara Falls has a temperature of 10.0°C. It falls a distance of 50 m. Assume that all of its potential energy is converted into thermal energy, calculate the temperature of the water at the bottom of the falls.

Solution:

$$\Delta T = \frac{E_{kin}}{c_{H_2O}\, m_{H_2O}} = \frac{m_{H_2O}\, g\, h}{c_{H_2O}\, m_{H_2O}} \qquad \textbf{(32)}$$

Which is independent of mass, so we can solve and find

$$\Delta T = \frac{g \cdot h}{c_{H_2O}} = 0.12^{\circ}C \qquad \textbf{(33)}$$

Problem 6.25

High speed stroboscopic photographs show that the head of a golf club of mass 200 g is travelling at 55 m/s as it strikes a golf ball of mass of 46 g. At the collision the golf ball is at rest on the tee. After the collision, the club head travels in the same direction at 40 m/s. Find the speed of the golf ball just after impact.

Solution: Write the conservation of momentum equations for the system that is the ball and the club. For illustration, we further use a slightly different notation with capital and small letters for the different velocities instead of indices 1 and 2:

$$M\, V_{initial} + m\, v_{initial} = M\, V_{final} + m\, v_{final} \qquad \textbf{(34)}$$

The capital letters refer to the club, the lowercase letters refer to the ball. Rearrange and solve for the speed of the golf ball just after impact:

$$v_{final} = \frac{M\, V_{initial} + m\, v_{initial} - M\, V_{final}}{m} \qquad \textbf{(35)}$$

Substituting the given data, this gives us v_{final} = 65 m/s. Given the ratio of the masses, this seems like a reasonable value.

CHAPTER SEVEN

Respiration: The properties of gases and cyclic processes

MULTIPLE CHOICE AND CONCEPTUAL QUESTIONS

Question 7.1
In Boyle's experiment, the volume of the sealed gas decreases as additional mercury is added to the open column. This effect is due to the fact that: (A) the volume of the gas decreases; (B) pressure and volume are linearly proportional to each other; (C) pressure and volume are proportional to each other; (D) pressure and volume are inversely proportional to each other; (E) pressure and volume are unrelated.

Answer: (D)

Question 7.3
We study the ideal gas equation. The gas constant can be given in different units, however, this unit is wrong:
(A) J/(K · mol); (B) (atm · m³)/(K · mol);
(C) (Pa · cm²)/(K · mol); (D) cal/(K · mol); (E) none of the above, all are suitable for the gas constant *R*.

Answer: (C)

Question 7.5
When food has been cooked in a pressure cooker it is very important to cool the container with cold water before removing the lid. Why?

Answer: The pressure inside exceeds atmospheric pressure while the contents are hot. If the lid is not secured, it will be accelerated off the pressure cooker.

Question 7.7
Organisms in the deep sea are subjected to very high pressures, as we will discuss in the context of Pascal's law in Chapter 11. Why are these organisms destroyed when they are pulled up to the surface? *Note*: An animal living at those depths is the giant squid. Nobody has yet seen a live specimen of this species!

Answer: We can model these organisms as a combination of liquids and gases, confined by their membrane or skin. If the external water pressure decreases rapidly, gases expand and may rupture the membrane or skin.

Question 7.9
In the kinetic gas theory developed by Boltzmann, Maxwell, and Clausius, which of the following is a result, not an assumption made to develop the model? (A) The gas consists of a very big number of particles with a combined volume that is negligible compared to the size of the container. (B) The internal energy of an ideal gas depends linearly on the temperature (in degrees kelvin). (C) The molecule size is much smaller than the inter–particle distance. (D) The molecules are in continuous random motion, travelling along straight lines while not colliding with other particles or the walls. (E) Collisions with each other and the walls of the box are elastic, which excludes intermolecular interactions.

Answer: (B)

Question 7.11
Which of the following statements about the temperature of a system is *not* correct? (A) The temperature of a single gas particle is determined by its velocity. (B) The temperature measurement in Fig. 6.32 is correct during the transition from liquid water to water vapour because the continuous addition of heat does not lead to a temperature change. (C) Temperature is a parameter which characterizes a system only when the system is in thermal equilibrium. (D) The temperature measurement with

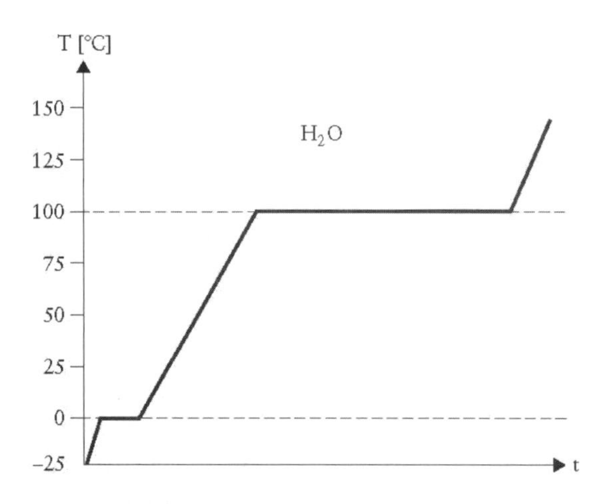

FIGURE 6.32

a Celsius thermometer is correct only when the expanding liquid in the thermometer and the system have the same temperature. (E) The temperature of the human body is usually higher than the air temperature in the immediate environment. Therefore, the human body and the surrounding air are not in thermal equilibrium.

Answer: (A)

Question 7.13
Why is heat required to boil a liquid even though the molecules in the liquid and the vapour share the same temperature?

Answer: Molecules attract each other (otherwise gases would never condense). Boiling leads to an increase in volume of the system, which means an increase of inter–molecular distances at a molecular scale. Energy is needed to separate molecules against this attractive force.

Question 7.15
Which of the following assumptions is *not* made in developing the kinetic gas theory? (A) The number of molecules is small. (B) The molecules obey Newton's law of motion. (C) Collisions between molecules are elastic. (D) The gas is a pure substance, not a mixture. (E) The average separation between molecules is large compared to their size.

Answer: (A)

Question 7.17
Small planets like Mercury and Mars have very thin or no atmospheres. Why?

Answer: Gases are held in a planetary atmosphere by gravity. Planets such as Mercury and Mars have smaller masses, and exert a smaller gravitational force. As the mass of a planet decreases, so does the minimum speed needed to escape from that planet's gravitational potential. The greater the gravitational force (due to planet mass) the greater the escape velocity for gas molecules, which is calculated from the kinetic energy needed to overcome gravity. Light molecules at typical planetary temperatures (particularly for the central region of the solar system) have a sufficient fraction of their Maxwell–Boltzmann distribution exceed the escape velocity; thus, the gases escape into outer space over the course of time. See also Problem 7.11.

Question 7.19
One container is filled with helium gas, another with argon gas. If both containers are at the same temperature, which molecules have the higher root–mean–square speed?

Answer: Helium, being the lighter gas, has the greater root–mean–square speed at the same temperature.

Question 7.21
A standard man inhales a tidal volume of air at 20°C. When the air arrives in the lungs, it has a temperature of 37°C. Treating air as an ideal gas, the change in internal energy is: (A) $\Delta U = 0$ J; (B) about an 85 % increase over the initial value; (C) about a 5 % increase over the initial value; (D) unknown, because the amount of air is not specified; (E) proportional to the change in the pressure of the air.

Answer: (C)

$$\frac{\Delta U}{U} = \frac{\frac{3}{2} n \cdot R\,(T_{final} - T_{initial})}{\frac{3}{2} n \cdot R\,T_{initial}} \qquad (1)$$

In this case, $T_{final} = 310$ K and $T_{initial} = 293$ K. When evaluated, this shows that the internal energy of the air increases by about 5% over the initial value.

Question 7.23
Fig. 6.21 shows a p–V diagram with three paths that a gas can take from an initial to a final state. Rank the paths in decreasing order according to (a) the change of internal energy ΔU, and (b) the amount of heat transfer Q between the system and the environment.

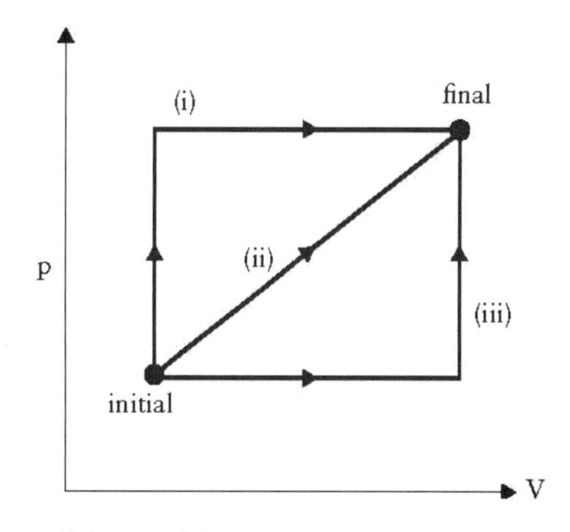

FIGURE 6.21

Answer to part (a): The same change occurs for the internal energy for all three paths because they have the same initial and final state.

Answer to part (b): (i) – (ii) – (iii). Heat transfer between the system and the environment is larger for paths that have changes in pressure occurring at higher temperatures. Since path (i) has all of its change in pressure occurring at high temperature, it has the most heat transfer between the system and the environment.

Question 7.25
We plot the following processes for 1 mol of an ideal gas in a p–V diagram. Which one leads to a linear plot, i.e., a plot which can be described by the linear formula $p = a + b \cdot V$? (A) An isothermal expansion, (B) an isobaric heating, or (C) an adiabatic cooling?

Answer: Definitively not (A) or (C). (B) is correct if we allow for $b = 0$.

Question 7.27
A process is an isochoric process when: (A) the temperature remains constant and the pressure changes; (B) no work is done; (C) no heat is exchanged with the environment; (D) the internal energy remains constant; (E) the pressure remains constant and the temperature changes.

Answer: (B). Note that non–isochoric reversible processes that return to the same state along the same path in a p–V diagram also have $W = 0$.

Question 7.29
A process is called adiabatic if: (A) the temperature remains constant, (B) no work is done, (C) no heat is exchanged with the environment, (D) the internal energy remains constant, (E) the volume remains constant.

Answer: (C)

Question 7.31
A standard man performs a strenuous exercise, e.g., lifting a weight or riding a bicycle. The standard man does work in this exercise and dissipates heat. Would the first law of thermodynamics not require that the internal energy, and therefore the temperature, of the standard man decreases?

Answer: The internal energy is diminished because both $W < 0$ and $Q < 0$ for the standard man as a closed system. However, the body temperature does not decrease as it is governed by the thermal energy, not the total energy of a system. Food energy is converted to thermal energy during the strenuous exercise.

Question 7.33
2.0 mol of an ideal gas is maintained at constant volume in a container of 4 L. If 100 J of heat is added to the gas, what is the change in its internal energy? (A) zero, (B) 50 J, (C) 67 J, (D) 100 J, (E) none of these values is correct.

Answer: (D). For an ideal gas undergoing a isochoric process, the change in internal energy is equal to the heat added to the system.

Question 7.35
We consider one breathing cycle, starting with an inhalation and then followed by an exhalation. The net work for such a cycle is (A) positive ($W > 0$), (B) zero ($W = 0$), or (C) negative ($W < 0$). *Note*: The gas in the lungs is the system.

Answer: (A). See Example 7.12.

Question 7.37
(a) We consider 1.0 litre of an ideal gas at STPD conditions. Describe a state into which this gas cannot be transferred.
(b) If you still need an ideal gas at the conditions you have chosen in part (a), how can you get it?

Answer to part (a): Same volume and temperature, but twice the pressure.

Answer to part (b): To double the pressure without changing the volume or temperature, you would have to double the amount of gas.

ANALYTICAL PROBLEMS

Problem 7.1
(a) Draw a graph for the volume of 1.0 mol of an ideal gas as a function of temperature in the range from 0 K to 400 K at constant gas pressures of, first, 0.2 atm and, second, 5 atm.
(b) Draw a graph for the pressure of 1.0 mol of an ideal gas as a function of volume between 0 L and 20 L at constant temperatures of 150 K and 300 K.

Solution to part (a): We use the ideal gas law in the form:

$$p \cdot V = n \cdot R \cdot T \qquad (2)$$

For the first plot, we need the volume to be the dependent variable and the temperature to be the independent variable:

$$V = \frac{n \cdot R}{p} T \qquad (3)$$

In this equation is $n = 1.0$ mol and the pressure is either $p = 0.2$ atm $= 20.26$ kPa, or $p = 5.0$ atm $= 506.5$ kPa. We plot in Fig. 7.44:

$$\begin{aligned} (I) \quad & V(m^3) = 4.1 \times 10^{-4}\ T(K) \\ (II) \quad & V(m^3) = 1.6 \times 10^{-5}\ T(K) \end{aligned} \qquad (4)$$

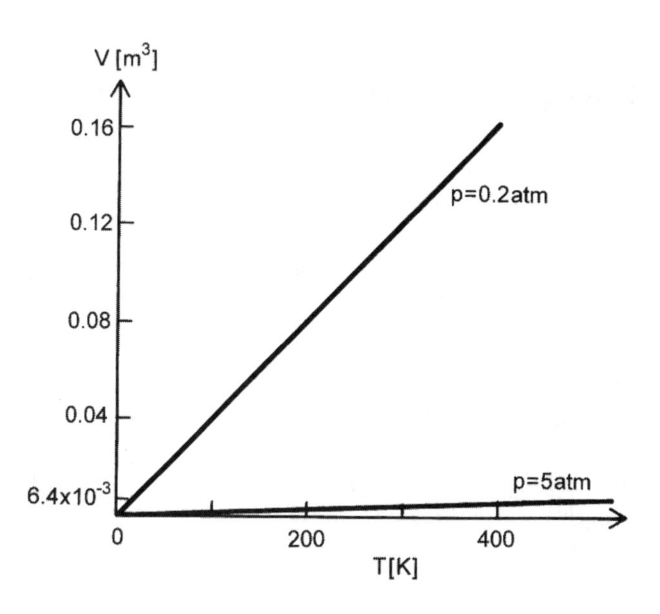

Figure 7.44

Solution to part (b): The ideal gas equation is rewritten such that the pressure is the dependent variable and the volume is the independent variable:

$$p = \frac{n \cdot R \cdot T}{V} \qquad (5)$$

Using again $n = 1.0$ mol, we plot Fig. 7.45:

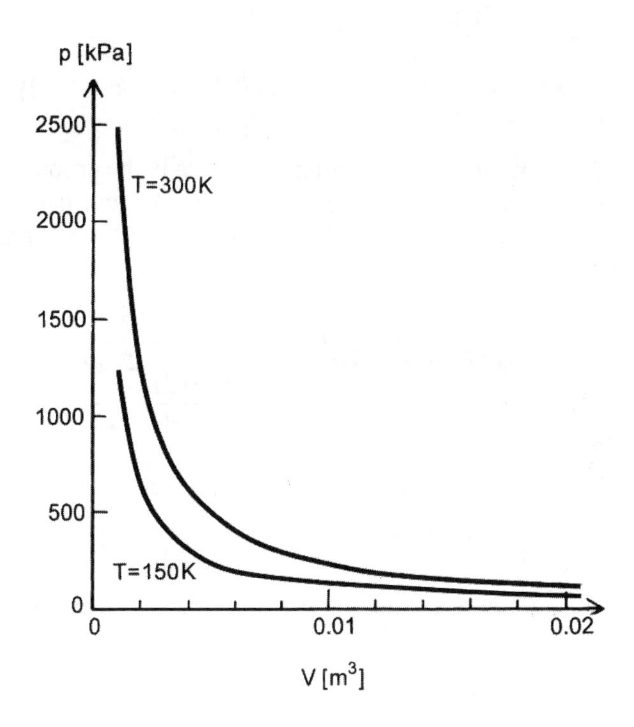

Figure 7.45

$$\begin{aligned} (I) \quad & T = 300\ K \Rightarrow p\,(kPa) = \frac{2.494}{V\,(m^3)} \\ (II) \quad & T = 150\ K \Rightarrow p\,(kPa) = \frac{1.247}{V\,(m^3)} \end{aligned} \qquad (6)$$

Problem 7.3

A container of volume $V = 400$ cm³ has a mass of 244.5500 g when evacuated. When the container is filled with air of pressure $p = 1$ atm at temperature $T = 20^0$C, the mass of the system increases to 245.0307 g. Assuming that air behaves like an ideal gas, calculate from these data the average molar mass of air.

Solution: This problem is an application of the ideal gas law. Since the mass of the gas is given and the molar mass is sought, we rewrite the ideal gas law in the form $p \cdot V = (m/M)\, R \cdot T$:

$$M = \frac{m \cdot R \cdot T}{p \cdot V} \qquad (7)$$

In Eq. [7], m is the mass of air, which is obtained from:

$$m = m_{full} - m_{evacuated} = 0.4807 \times 10^{-3}\ kg \qquad \textbf{(8)}$$

By substituting the remaining data given in the problem, including a temperature of 20^0C = 293 K, and the air pres-sure in the container as 1.0 atm = 1.013×10^5 Pa, we find:

$$M = \frac{0.4807 \times 10^{-3}\ kg \left(8.314\ \frac{J}{K\ mol}\right) 293\ K}{(1.013 \times 10^5\ Pa)\,(400 \times 10^{-6}\ m^3)} \qquad \textbf{(9)}$$

$$= 0.0289\ \frac{kg}{mol} = 28.9\ \frac{g}{mol}$$

We usually use a value of 29 g/mol for the molar mass of air.

Problem 7.5
The pressure of an ideal gas is reduced by 50%, resulting in a decrease in temperature to 75% of the initial value. Calculate the ratio of final to initial volumes of the gas.

<u>Solution:</u> $V_{final}/V_{initial} = 3/2$ because

$$\frac{V_2}{V_1} = \frac{p_1}{p_2}\frac{T_2}{T_1} = \left(\frac{2}{1}\right)\left(\frac{3}{4}\right) = \frac{3}{2} \qquad \textbf{(10)}$$

Problem 7.7
An ideal gas is confined in a container at a pressure of 10.0 atm and at a temperature of 15^0C. If 50% of the gas leaks from the container and the temperature of the re-maining gas rises to 65^0C, what is the final pressure in the container?

<u>Solution:</u> p_{final} = 5.87 atm because

$$p_2 = \frac{n_2}{n_1}\frac{T_2}{T_1}p_1 = 5.87\ atm \qquad \textbf{(11)}$$

Be sure to convert the temperatures into kelvin.

Problem 7.9
A spherical weather balloon is designed to inflate to a maximum diameter of 40 m at its working altitude, where the air pressure is 0.3 atm and the temperature is 200 K. If the balloon is filled at atmospheric pressure and temperature 300 K, what is its radius at lift–off? Treat the gas as an ideal gas.

<u>Solution:</u> Use the ideal gas law to find:

$$r_1 = \left(\frac{T_1}{T_2}\frac{p_2}{p_1}r_2^3\right)^{1/3} = 15.3\ m \qquad \textbf{(12)}$$

Note that the subscript 1 refers to the state on the ground, and the subscript 2 refers to the state in the air. Also, recognize that what is given in the question is a dia-meter, when what we use in this formula is the radius.

Problem 7.11
The temperature in the upper regions of the atmosphere of Venus is 240 K.
(a) Find the root–mean–square speed of hydrogen mo-lecules (H_2) and carbon–dioxide (CO_2) in that region.
(b) A result of planetary science is that a gas eventually is lost from a planet's atmosphere into outer space if its root–mean–square speed is one–sixth of the escape velo-city, which can be calculated from gravity. Using an escape velocity of 10.3 km/s for Venus, does either of the two gases in part (a) escape from that planet?

<u>Solution to part (a)</u>: Use

$$v_{rms} = \sqrt{\frac{3RT}{M}} \qquad \textbf{(13)}$$

with M_{H2} = 0.002 kg/mol and M_{CO2} = 0.044 kg/mol to find $v_{rms}(H_2)$ = 1.73 km/s; $v_{rms}(CO_2)$ = 370 m/s.

<u>Solution to part (b)</u>: $v_{escape}/6$ = 1.71 km/s. Since $v_{rms}(H_2)$ is slightly larger than this, it will eventually be depleted from Venus.

Problem 7.13
1.0 mol of an ideal gas isothermally expands from an ini-tial pressure of 20 atm to a final pressure of 5 atm. Cal-culate separately for two temperatures, 0^0C and 25^0C,
(a) the work done by the gas,
(b) the change of internal energy of the gas, and
(c) the amount of heat taken from the environment.

<u>Solution to part (a)</u>: This problem requires that the same type of calculation is done twice, once for an isothermal expansion at $T_1 = 0^0C = 273$ K and once for an iso-thermal expansion at $T_2 = 25^0C = 298$ K.

For the work of an isothermal expansion, we use the formula developed in Eq. [7.100] with the pressure as independent variable instead of the volume. We find at T_1:

$$W_1 = -n \cdot R \cdot T_1 \ln\left(\frac{p_{initial}}{p_{final}}\right)$$

$$= -(1\ mol)\left(8.314\ \frac{J}{K\ mol}\right)(273\ K)\ln\left(\frac{20}{5}\right) \quad \textbf{(14)}$$

$$= -3.15\ kJ$$

At T_2 we obtain $W_2 = -3.45$ kJ. This result can also be obtained by calculating:

$$W_2 = W_1 \frac{T_2}{T_1} \quad \textbf{(15)}$$

Solution to part (b): The change of the internal energy of an ideal gas during an isothermal process is zero, i.e., $\Delta U = 0$ since $\Delta T = 0$. This applies at any temperature.

Solution to part (c): Using the first law of thermodynamics for closed systems, we obtain:

$$Q = \Delta U - W = -W \quad \Rightarrow$$

$$Q_1 = +3.15\ kJ \quad ; \quad Q_2 = +3.45\ kJ \quad \textbf{(16)}$$

Problem 7.15

Show that Poisson's equation leads to $p \cdot V^{\kappa} = const$ when using the operations specified in the text.

Solution: We start with the equation derived in the text, Eq. [7.115]. In this formula we replace the temperature by $p \cdot V/(n \cdot R)$ from the ideal gas law:

$$V \cdot T^{C_V/R} = V\left(\frac{p \cdot V}{n \cdot R}\right)^{C_V/R} = const$$

$$p^{\frac{C_V}{R}} \cdot V^{\frac{C_V}{R}+1} = const\ (n \cdot R)^{C_V/R} = const^* \quad \textbf{(17)}$$

in which const* is indeed a constant since both n and R in the middle term of the second formula in Eq. [17] do not vary. In the next step we rewrite the exponent of the volume term:

$$p^{\frac{C_V}{R}} \cdot V^{\frac{C_V}{R}+1} = p^{\frac{C_V}{R}} \cdot V^{\frac{C_V+R}{R}} = const^* \quad \textbf{(18)}$$

In the last step, we raise both sides of the formula to the (R/C_V)–th power:

$$p \cdot V^{\frac{C_V+R}{C_V}} = \left(const^*\right)^{\frac{R}{C_V}} = const' \quad \textbf{(19)}$$

in which const' is also constant since the middle term in Eq. [19] does not contain any variables. Eq. [19] is easily transformed into Eq. [7.116] when using the relation between C_V and C_p and yields $\kappa = C_p/C_V$.

Problem 7.17

In the text we is stated that $C_p = C_V + R$ holds for an ideal gas. Derive this result for an isobaric expansion of 1.0 mol of ideal gas without using the enthalpy concept. For this, start with the work in Eq. [7.102] and the change of internal energy in Eq. [7.76] for the expansion in Fig. 7.31. Then use the first law of thermodynamics and the definition

$$Q = n \cdot C_p \cdot \Delta T \quad \textbf{(20)}$$

which applies for an isobaric process.

Solution: We rewrite the first law of thermodynamics, $\Delta U = W + Q$ in the form $W = -n \cdot p \cdot \Delta V$. The change of the internal energy is $\Delta U = n \cdot C_V \cdot \Delta T$. We further use Eq. [20]:

$$\Delta U = n \cdot C_V \cdot \Delta T = -n \cdot p \cdot \Delta V + n \cdot C_p \cdot \Delta T \quad \textbf{(21)}$$

which leads to:

$$C_p - C_V = \frac{n \cdot p \cdot \Delta V}{n \cdot \Delta T} = \frac{p \cdot \Delta V}{\Delta T} = R \quad \textbf{(22)}$$

Problem 7.19

Sketch a p–V diagram for the following processes:

(a) An ideal gas expands at constant pressure p_1 from volume V_1 to volume V_2; is then kept at constant volume while the pressure is reduced to p_2.

(b) An ideal gas is reduced in pressure from p_1 to p_2 while its volume is held constant at V_1. It is then expanded at constant pressure to a final volume V_2.

(c) In which process is more work done?

Solution to part (a): See Fig. 7.46(a)

Solution to part (b): See Fig. 7.46(b)

Solution to part (c): Identifying the gas the system, we find $W(\text{a}) < W(\text{b}) < 0$.

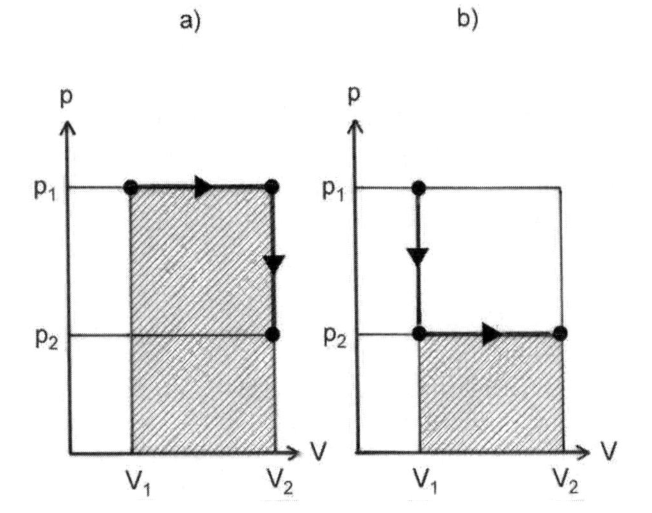

Figure 7.46

Problem 7.21

1.0 mol of an ideal gas is initially at 0°C. It undergoes an isobaric expansion at $p = 1.0$ atm to four times its initial volume.

(a) Calculate the final temperature of the gas, T_{final}.

(b) Calculate the work done by/on the gas during the expansion.

Solution to part (a):

$$T_2 = \frac{p V_2}{n R} = \frac{p\, 4\, V_1}{n R} = \frac{4 p \left(\dfrac{n R T_1}{p} \right)}{n R} \qquad \textbf{(23)}$$

$$= 4\, T_1 = 1090\ K$$

Solution to part (b):

$$W = -p\,(V_{final} - V_{initial}) = -6.81\ kJ \qquad \textbf{(24)}$$

Problem 7.23

Compare the efficiency coefficient for a Carnot machine operating between a low–temperature heat reservoir at room temperature (25°C) and a high–temperature heat reservoir at the boiling point of water at two different pressures:

(a) 5 atm with $T_{\text{boil}} = 152^0$C, and

(b) 100 atm with $T_{\text{boil}} = 312^0$C.

Solution to part (a): This is a straight–forward substitution problem for the efficiency coefficient of the Carnot process; at pressure 5 atm we find:

$$\eta = \frac{T_b - T_a}{T_b} = \frac{425\ K - 298\ K}{425\ K} = 30\% \qquad \textbf{(25)}$$

Solution to part (b): At pressure 100 atm, we find η = 49%. It is important to notice, however, that the higher efficiency coefficient is not a result of the changed pressure, but follows only indirectly from the pressure change as a boiling point change is involved.

Problem 7.25

The cyclic process in Fig. 7.42 consists of (i) an isothermal expansion, (ii) an isochoric cooling, and (iii) an adiabatic compression. If the process is done with n = 1.0 mol of an ideal gas, what is

(a) the total work done by the gas,

(b) the heat exchanged with the environment, and

(c) the change of the internal energy for one cycle.

(d) Sketch the cyclic process of Fig. 7.42 as p–T, V–T and U–T diagrams.

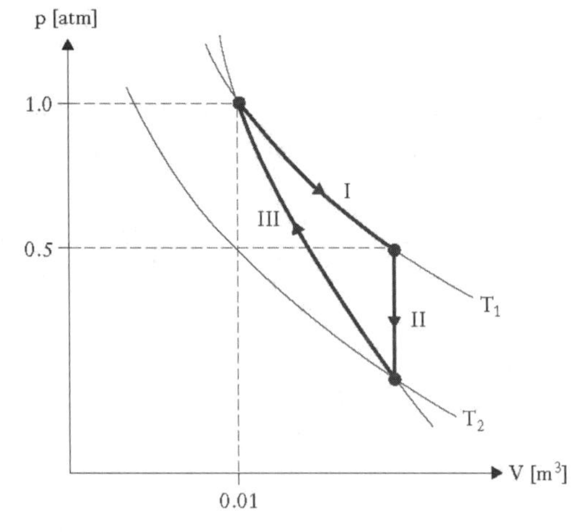

FIGURE 7.42

Solution to part (a): Any cyclic process of this type is treated the same way that we dealt with the Carnot process. In the first step we summarize the work, heat and internal energy contributions due to the single steps in-

Table 7.3

Process step	work W	heat Q
I	$-n\,R\,T_1\,\ln(p_{initial}/p_{final})$	$-W_I$
II	0	$-(3/2)\,n\,R\,(T_1-T_2)$
III	$+(3/2)\,n\,R\,(T_1-T_2)$	0

volved in the cyclic process. For this we need not to calculate any of the contributions since each individual step has been covered in Chapter 7. The contributions are summarized in Table 7.3. To quantify the terms in Table 7.3, the temperatures of the two isothermal curves in Fig. 7.42 must be calculated. The temperature T_1 is obtained from the ideal gas law applied to the first state of the system:

$$T_1 = \frac{p_1 \cdot V_1}{n \cdot R}$$

$$= \frac{1.013 \times 10^5\,Pa)\,(0.01\,m^3)}{\left(8.314\,\frac{J}{K\,mol}\right)\,1.0\,mol} = 121.8\,K \quad \textbf{(26)}$$

The second temperature is obtained from the adiabatic step (step III) in Fig. 7.42, using $C_V/R = 3/2$ for an ideal gas:

$$T_2 = \left(\frac{V_1}{V_2}\right)^{2/3} T_1 = 0.5^{2/3}\,(121.8\,K) = 76.8\,K \quad \textbf{(27)}$$

in which the ratio of the two volumes is 0.5 because of Boyle's law applied to step (I) in Fig. 7.42: $p_2/p_1 = V_1/V_2$. The total work is the sum of the three work terms in Table 7.3:

$$W_I = -1.0\,mol\left(8.314\,\frac{J}{Kmol}\right)(121.8\,K)\ln\frac{1.0}{0.5}$$
$$= -700\,J$$

$$\textbf{(28)}$$

$$W_{III} = +\frac{3}{2}\,1.0\,mol\left(8.314\,\frac{J}{K\,mol}\right)45\,K$$
$$= +560\,J$$

in which $T_1 - T_2 = 121.8\text{ K} - 76.8\text{ K} = 45.0\text{ K}$ is the temperature difference in the third work term. Thus, the work per cycle is $W_{cycle} = W_I + W_{III} = -140$ J. Because this is a negative value, the work is done by the gas.

<u>Solution to part (b)</u>: From Table 7.3, we note that $Q_{II} = -W_{III}$ and $Q_I = -W_I$. Thus, $Q_{cycle} = -W_{cycle} = +140$ J. Since the value is positive, we conclude that the system receives this amount of energy as heat per cycle.

<u>Solution to part (c)</u>: For any cyclic process the change per cycle in the internal energy must be zero because the internal energy is a variable of state of the system. This can be confirmed by adding Q_{cycle} and W_{cycle}, which were calculated above.

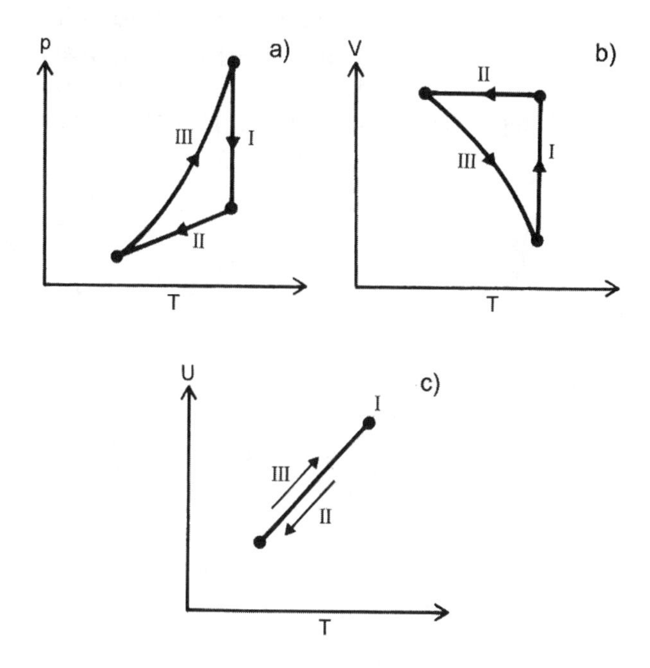

Figure 7.47

<u>Solution to part (d)</u>: The three graphs are shown in Fig. 7.47. They are derived in the following fashion:

- *p–T plot*: The first step is an isothermal expansion or isothermal decompression. Since temperature is constant for isothermal processes, the initial and final state must lie on a common, vertical line. The second step is an isochoric cooling, i.e., both, temperature and pressure decrease while the volume stays constant. The third step must bring the system back to the initial state.
- *V–T plot*: The first step is an isothermal expansion, i.e., the volume increases and the temperature remains the same (vertical line). The second step is an isochoric cooling, i.e., the temperature is reduced while the volume is constant (horizontal line). The last step must connect back to the initial state. Both curves for the last step are bent because the adiabatic process is not a linear relation between either pressure and temperature or volume and temperature.
- *U–T plot*: Since the first step is isothermal, both the temperature and the internal energy of the ideal gas do not change. Thus, during the first step the system does not change its position in this type of plot. The second step is an isothermal cooling, i.e., the temperature de-

creases, and with it the internal energy. This relationship is linear due to $\Delta U \propto \Delta T$. The third step must bring the system back to the initial state.

Problem 7.27

Fig. 7.41 shows the p–V relationship in the left ventricle of the human heart. The curve is traversed counter–clockwise with increasing time. Using the stroke volume and pressure data from Question 7.34, determine graphically the amount of work done in a single cycle. *Hint*: Simplify the calculation by using the dashed straight lines in the *p–V* diagram instead of curved segments.

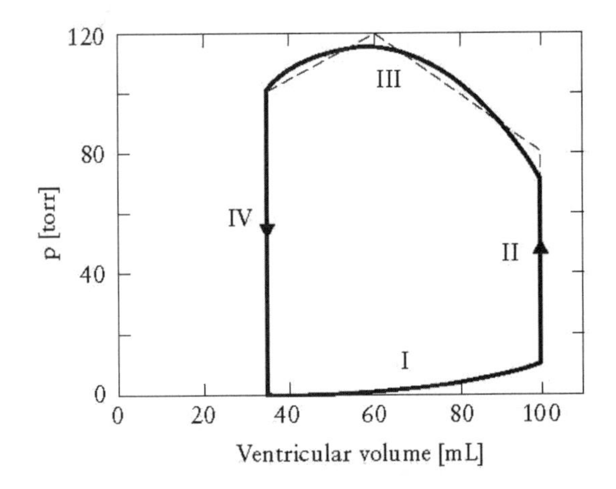

FIGURE 7.41

<u>*Solution:*</u> Work is defined as the area under the curve in a *p–V* diagram. To conveniently analyse the area under the curve in Fig. 7.41 (using the dashed lines instead of the curved upper line), the area is subdivided into six sections, as shown in Fig. 7.48. Note that sections 1 to 5 contribute to the area under the curve, which is traversed from larger toward smaller volume, and thus, lead to positive work contributions. Section 6 represents an area which must be subtracted from the combined areas 1 through 5 to properly describe the enclosed area. Note that all six areas are either rectangular or triangular with a 90^0 angle. The area of these two shapes are one or one–half times the width times the height, respectively. Thus, we find (all values given in unit torr · mL):

Area 1 = ½ (25 · 20) = 250, Area 2 = ½ (40 · 40) = 800, Area 3 = (100 · 25) = 2500, Area 4 = (70 · 40) = 2800, Area 5 = (10 · 40) = 400, and Area 6 = ½ (10 · 40) = 200.

We obtain for the work from these data: W = 6750 torr · mL – 200 torr · mL = 6550 torr · mL. This is converted to standard unit with 1 mL = 1×10^{-6} m^3 and 1 torr = 133.32 Pa, yielding a work of $W = +\ 0.87$ J.

The work is positive, i.e., work is done on the fluid in the heart (here not an ideal gas but the liquid blood).

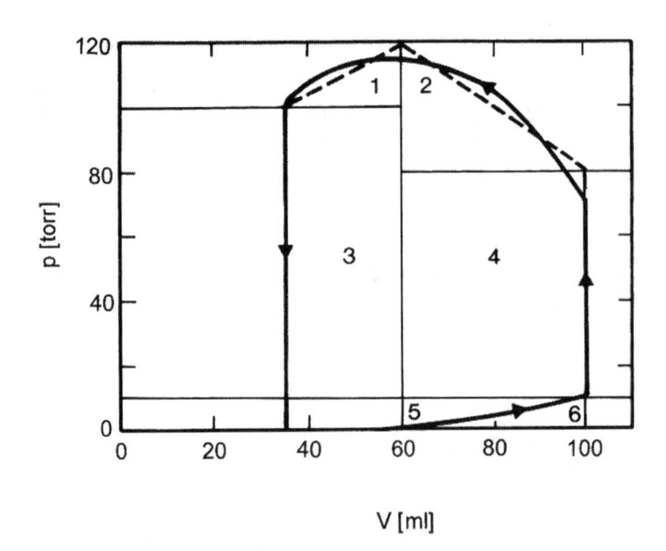

Figure 7.48

CHAPTER EIGHT

Molecular biology: Basic physics for biochemistry

MULTIPLE CHOICE AND CONCEPTUAL QUESTIONS

Question 8.1

1 mol of an ideal gas is expanded from initial volume $V_{initial}$ to a final volume V_{final}, in each case starting at pressure p_0.

(a) For which type of reversible process does the gas do the most work? (A) adiabatic expansion, (B) isothermal expansion, (C) isobaric expansion, (D) cannot be determined without further information?

(b) For which type of irreversible process does the gas do the most work: (A) adiabatic expansion, (B) isothermal expansion, (C) isobaric expansion, (D) cannot be determined without further information.

(c) If you chose one of the three processes in part (a), explain why the other two are still reversible despite the definition that a reversible process is the process that involves the most work.

Answer to part (a): (C)

Answer to part (b): (D)

Answer to part (c): The reversible process for a given process is the one in which the most work is done. The comparison is only between equivalent processes, not between processes that follow different paths in a *p–V* diagram for the various variables of the state.

Question 8.3

Marcellin Berthelot measured a large number of standard enthalpy of formation values. He justified that effort in an 1878 publication by saying that "every chemical process in an isolated system tends toward the products that release the most heat"; i.e., Berthelot believed that his measurements reveal the chemical affinity of the reactants. Do you agree?

Answer: Berthelot stated this in his publication "Essai de Mécanique chimique." Were Berthelot correct, no endothermic reaction could occur spontaneously. Also the reversibility of most chemical reactions would not be possible. To understand chemical affinity, the entropy of the second law of thermodynamics must be included.

Question 8.5

The following statement about the entropy of a system is not correct: (A) It is a measure of the degree of order/disorder in the system. (B) It is a measure of the fraction of heat taken up by the system that is turned into work during a reversible cyclic process. (C) It determines whether an isolated system is in equilibrium or may undergo spontaneous processes. (D) It can be combined with the temperature and the enthalpy to determine the Gibbs free energy of the system. (E) The entropy difference between two states of the system is equal to the amount of heat exchanged with the environment during a process leading from one state to the other.

Answer: (E)

Question 8.7

A typical 1800s steam engine operated with steam of 125^{0}C. Room–temperature air served as the low–temperature heat reservoir (T = 25^{0}C). What was the efficiency coefficient η of that machine? (Choose the closest value.) (A) 100%, (B) 80%, (C) 34%, (D) 25%, (E) 10%.

Answer: (D):

$$\eta = \frac{T_{high} - T_{low}}{T_{high}} = \frac{398\,K - 298\,K}{398\,K} = 25\% \quad \textbf{(1)}$$

Question 8.9

A machine operates within an isolated superstructure. Its operation causes an entropy increase of ΔS = 5 J/K. What change ΔS^* is required in the machine's environment within the isolated superstructure to identify the machine as operating reversibly? (A) ΔS^* = 0 J/K; (B) ΔS^* = + 5 J/K; (C) ΔS^* = – 5 J/K; (D) no value exists to answer the question; (E) the value of ΔS is different from those listed above.

Answer: (C). A reversible process is characterized by a zero change in entropy for the isolated superstructure. That means that the system entropy when added to the environment entropy must equal zero.

Question 8.11
A piece of ice at temperature 0^0C and 1.0 atm pressure has a mass of 1.0 kg. It then melts completely to water. What is its change in entropy? The latent heat of freezing of water is given as 3.34×10^5 J/kg. (Choose the closest value.) (A) 3340 J/K. (B) 2170 J/K. (C) 613 J/K. (D) 1220 J/K.

Answer: (D). The latent heat of freezing of water is 3.34×10^5 J/kg, and we have 1.0 kg of ice. This gives us $\Delta H = 3.34 \times 10^5$ J. $\Delta S = \Delta H/T = 3.34 \times 10^5$ J / 273 K. Which gives us a change in entropy of 1220 J/K. Note that the latent heat of freezing is written as an enthalpy because the process occurs at constant pressure, not at constant volume.

Question 8.13
A thermodynamic process occurs in which the entropy of a system changes by – 8.0 J/K. Based on the second law of thermodynamics, what can you state about the entropy change in the environment?

Answer: $\Delta S_{environment} \geq +8.0$ J/K.

ANALYTICAL PROBLEMS

Problem 8.1
We revisit the Carnot process discussed in Problem 7.24. This process is operated with 1.0 mol of an ideal gas of heat capacity $C_V = 3 \cdot R/2$. The pressure of the gas is 10.0 atm and the temperature is 600 K in the most compressed state. From there, an isothermal expansion leads to a pressure of 1.0 atm. The lower process temperature is 300 K. Draw this Carnot process in an *S*–*T* diagram.

Solution: We note the changes in parameter values to Problem 7.24. The resulting plot is shown in Fig. 8.14. Note that it is a cyclic process. Note also that *adiabatic* as a term also describes a reversible *isentropic* (no entropy change) process.

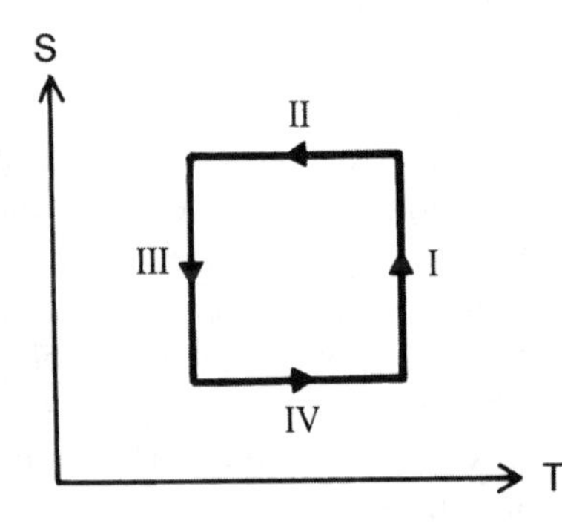

Figure 8.14

Problem 8.3
Calculate the entropy change during melting of 1.0 mol of benzene. The melting point of benzene at atmospheric pressure is $T_{melt} = 288.6$ K and its latent heat of melting is 30 kcal/kg.

Solution: We use the definition of the entropy change for the process, $\Delta S = Q/T$. To express the entropy in unit J/(K · mol), we need to convert the latent heat to a molar value. The molar mass of benzene is:

$$M_{C_6H_6} = 6\left(M_C + M_H\right) = 6\left(12.01\ \frac{g}{mol} + 1.01\ \frac{g}{mol}\right) = 78.1\ \frac{g}{mol} \quad \textbf{(2)}$$

With this mass, we convert the latent heat of melting, which is 30 kcal/mol = 126 kJ/mol:

$$\Delta H_{melt} = \left(126\ \frac{kJ}{kg}\right)\left(78.1\ \frac{g}{mol}\right) = 9.84\ \frac{kJ}{mol} \quad \textbf{(3)}$$

Note that the latent heat of melting is written as an enthalpy because the process occurs at constant pressure, not at constant volume. With the melting temperature T_{melt}, we find:

$$\Delta S_{melt} = \frac{\Delta H_{melt}}{T_{melt}} = \frac{9.84\ \frac{kJ}{mol}}{288.6\ K} = 34.1\ \frac{J}{K\ mol} \quad \textbf{(4)}$$

Problem 8.5
We place 150 g ice at 0^0C (latent heat of melting is 1430 cal/mol) in a calorimeter with 250 g water at 80^0C. Use 18.0 cal/(K · mol) for the molar heat capacity of liquid water and assume that this value is temperature–independent.
(a) What is the final temperature of the water?
(b) If the process is done reversibly, what is the entropy change of the combined system ice/water?
(c) What is the entropy change in the environment for the reversible process?
(d) What is the entropy change if the process is done irreversibly in an isolated beaker?
Hint: Use Eq. [8.28] for the temperature dependence of the entropy:

$$\Delta S = n \cdot C_p \ln\left(\frac{T_{final}}{T_{initial}}\right) \quad \textbf{(5)}$$

Solution to part (a): We first calculate the amount of energy needed to melt 150 g of ice. This energy must be extracted from the water at 80^{0}C, leaving that part of the system at a lower temperature. In the last step, we mix the 150 g water now liquid at 0^{0}C with the 250 g of warmer water.

The energy needed to melt 150 g of water is obtained from the given latent heat of melting and the molar mass of water, M = 18 g/mol:

$$E_{melt} = \frac{5.98 \frac{kJ}{mol}}{18 \frac{g}{mol}} 150\ g = 49.9\ kJ \quad \textbf{(6)}$$

The same energy is obtained from cooling 250 g of water from $T_{initial}$ = 80^{0}C = 353 K to T_{final}. Because the process is done at constant pressure, we use C_p in:

$$\Delta H = -n \cdot C_p (T_{final} - T_{initial}) \quad \textbf{(7)}$$

which yields for the final temperature:

$$T_{final} = -\frac{\Delta H}{n \cdot C_p} + T_{initial} \quad \textbf{(8)}$$

and with the given values substituted:

$$T_{final} = -\frac{49.9\ kJ}{\frac{250\ g}{18\ g/mol}\left(75.3 \frac{J}{K\ mol}\right)} + 363\ K \quad \textbf{(9)}$$

$$= 315.3\ K = 42.2^0C$$

Mixing 150 g water at 0^{0}C and 250 g of water at 42.2^{0}C leads to 400 g of water at 26.4^{0}C.

Solution to part (b): The entropy change is determined in two parts: the contribution of the phase change of ice and the contribution due to the temperature changes in the liquid water. The first contribution to the entropy change is the melting of 150 g ice. We saw that the energy needed for that process is 49.9 kJ. The associated entropy value is ΔS = + 182.6 J/K because the melting happens at 273.1 K.

Eq. [5] is now applied for each of the two amounts of water in the system, for 150 g to heat from the freezing point to 26.4^{0}C, and for 250 g of water to cool from 80^{0}C to 26.4^{0}C. Note that the heating leads to a positive entropy change and the cooling to a negative change; thus ΔS is written as:

$$\frac{150\ g}{18 \frac{g}{mol}}\left(75.3 \frac{J}{K\ mol}\right)\left(\ln\frac{299.5}{273.1} + \frac{2.5}{1.5}\ln\frac{299.5}{353.1}\right) \quad \textbf{(10)}$$

The factor before the last logarithm–term of Eq. [10] recognizes that 250 g water are cooled, not just 150 g. Eq. [10] leads to ΔS = – 114.3 J/K.

We combine both contributions to the entropy change of the system:

$$\Delta S_{system} = 182.6 \frac{J}{K} - 114.3 \frac{J}{K} = +68.3 \frac{J}{K} \quad \textbf{(11)}$$

Solution to part (c): For the process to be reversible, the change in the environment must be equal but opposite to the change in the system. Thus, $\Delta S_{environment}$ = – 68.3 J/K.

Solution tp part (d): Here we get ΔS_{system} = + 68.3 J/K and $\Delta S_{environment}$ = 0 J/K; i.e., the total entropy change is an increase of 68.3 J/K. This is a thermodynamically possible process.

Problem 8.7

Determine graphically the standard entropy of silver from the data given in Table 8.8.

TABLE 8.8

Molar heat capacity C_p for silver (Ag) at various temperatures

T (°C)	C_p (cal/(K · mol))	T (°C)	C_p (cal/(K · mol))	T (°C)	C_p (cal/(K · mol))
−263	0.07	−243	1.14	−223	2.78
−203	3.90	−183	4.57	−163	5.01
−143	5.29	−123	5.49	−103	5.64
−83	5.76	−63	5.84	−43	5.91
−23	5.98	−3	6.05	+17	6.08

Solution: The standard entropy for 1 mol is determined from the definition:

$$\Delta S = S_{final} - S_{initial} = \sum_i \frac{Q_i}{T_i} \quad \textbf{(12)}$$

We choose the initial state to be when the temperature is 0 K, at which $S = 0$ J/K. We choose the final state as the standard state at temperature 298 K. To obtain a value at constant pressure, the heat term in Eq. [12] is replaced by the term $Q = C_p \cdot \Delta T$:

$$S^0 = \sum_{T=0K}^{298K} \frac{C_p}{T} \Delta T \quad \textbf{(13)}$$

We obtain the standard entropy from the data in Table 8.8 by plotting C_p/T versus the temperature. This plot is shown in Fig. 8.15.

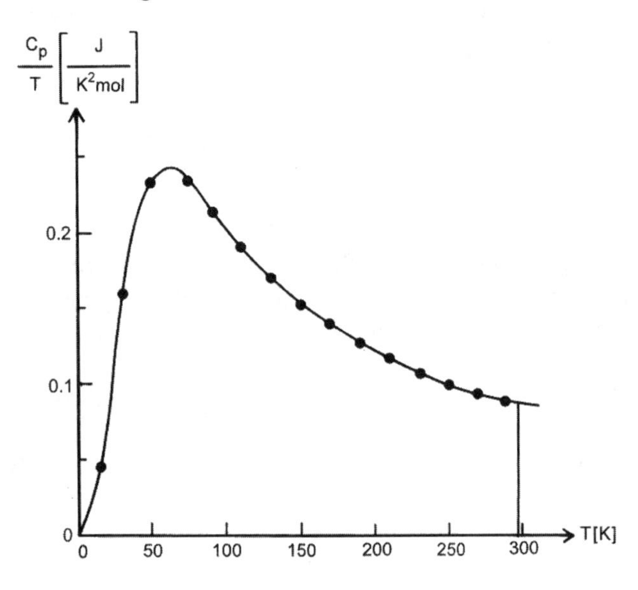

Figure 8.15

Problem 8.9
Derive Eq. [8.4] from Eq. [8.3].

Solution: We want to show how we get from the formula for the efficiency coefficient of the Carnot process

$$\eta = \frac{|W|}{Q_b} = \frac{Q_a + Q_b}{Q_b} = \frac{T_b - T_a}{T_b} \quad \textbf{(14)}$$

to the formula which motivates the introduction of the entropy:

$$\frac{Q_a}{T_a} + \frac{Q_b}{T_b} = 0 \quad \textbf{(15)}$$

We start with the last two terms in Eq. [14]. Eliminating the denominators through multiplication yields:

$$Q_a T_b + Q_b T_b = Q_b T_b - Q_b T_a \quad \textbf{(16)}$$

which is equal to:

$$Q_a T_b = - Q_b T_a \quad \textbf{(17)}$$

Eq. [17] is divided on both sides by the product $T_a \cdot T_b$:

$$\frac{Q_a}{T_a} = - \frac{Q_b}{T_b} \quad \textbf{(18)}$$

This is equivalent to Eq. [15].

Problem 8.11. Calculate ΔS and ΔG for the evaporation of 1 mol of water at $T = 100^0$C and $p = 1$ atm. The latent heat of evaporation of water is 9.7 kcal/mol.

Solution: We start with the entropy change. We use $n = 1$ mol, $T = 373$ K, and convert the latent heat unit:

$$\Delta H_{evaporation} = 9.7 \frac{kcal}{mol} = 40.6 \frac{kJ}{mol} \quad \textbf{(19)}$$

with these values, we find:

$$\Delta S_{evaporation} = \frac{\Delta H_{evaporation}}{T} = \frac{40.6 \frac{kJ}{mol}}{373\,K} = 108.8 \frac{J}{K \cdot mol} \quad \textbf{(20)}$$

We note that water evaporation at 100^0C (p = 1 atm) occurs between two states in thermal equilibrium. Thus, we expect at $T = T_{evaporation}$ that $\Delta G = 0$:

$$\Delta G_{evaporation} = \Delta H_{evaporation} - T \cdot \Delta S_{evaporation} = \Delta H_{evaporation} - T \frac{\Delta H_{evaporation}}{T_{evaporation}} = 0 \quad \textbf{(21)}$$

CHAPTER NINE

Blood and Air: Mixed phases

MULTIPLE CHOICE AND CONCEPTUAL QUESTIONS

Question 9.1

Extracellular fluids and the cytoplasm of most human cells have a concentration of osmotically active components of 0.29 mol/kg, due to dissolved sugar or ions such as potassium, sodium, and chlorine.
(a) If the body loses water (dehydration) or has a too–large intake of salt, both of which affect the extracellular fluid, what consequence does this have for the body cells?
(b) What would happen if, instead, a human body were to lose too much salt from the extracellular fluid?

Answer to part (a): We first consider the case in which the concentration of the active osmotic components in the extracellular fluid is higher than in the cells. As a consequence, water diffuses from inside the cell toward the extracellular space. The reason for this diffusion is given in Eqs. [9.42] and [9.46]:

$$\Pi = p_{mix} - p_{pure} = \frac{RT}{V_1^0} x_2 \qquad (1)$$

This formula applies at both sides of the membrane, and thus, establishes a pressure gradient for the solvent across the membrane. An increase in the salt concentration outside of the cell membrane means that a larger pressure difference is required for diffusive equilibrium. Since the extracellular pressure does not change, the cytoplasm pressure must decrease, which is achieved by water diffusion out of the cell. This dehydration of the cell leads to cell shrinkage which may not be reversible, i.e., it leads to permanent damage to the cell.

Answer to part (b): This is the opposite situation to part (a). In this case, Eq. [1] predicts a smaller pressure difference in diffusive equilibrium. Thus, water diffuses from the extracellular space into the cell, leading to cell swelling. This may cause the cell to burst, which is a form of permanent tissue damage.

Question 9.3

In 1965, a French team led by Jacques–Yves Cousteau lived for 28 days in the deep–sea station *Conshelf III* at 108 m below the sea surface. They breathed an oxygen/helium mixture (called heliox) instead of air.
(a) Would you agree with their claim that breathing this mixture is easier than breathing air?
(b) Can you think of another reason why they used heliox instead of air? *Note*: One member of the team reported that among other adverse effects of this exercise irritated his tastebuds, and he could no longer distinguish caviar from chicken. This, of course, is a disastrous effect for a Frenchman!

Answer to part (a): This question jumps a bit ahead as we discuss the transition from laminar to turbulent flow based on Osbourne Reynolds model in Chapter 12. Reynolds showed that turbulent flow is favoured for fluids of large gas density ρ (see Eq. [12.64]). The gas density is reduced when the inert component of air (nitrogen) is replaced by helium. The benefit of this gas exchange is a reduction of the likelihood that the gas mixture flows turbulently through the trachea, thus, making breathing easier.

Answer to part (b): The original reason for the exchange of helium for nitrogen was *nitrogen narcosis*. In earlier deep sea diving experiments, Cousteau's team had noted that a dangerous narcotic effect occurs at depths below 45 metres due to the high partial pressure of nitrogen. For example, Cousteau reported that this effect caused one team member to offer his mouth–piece to fishes passing–by. Replacing nitrogen with helium eliminated this effect.

Besides the irritation of taste, there are other interesting changes taking place when heliox is used instead of air:
(I) A funny change in the sound of the human voice takes place, causing people to sound like Mickey Mouse. This effect is also sometimes used by entertainers at children parties. It was considered an undesirable effect when the Cousteau team went on a live TV show. Therefore, on that occasion, heliox was replaced by an expensive neon/oxygen mixture.
(II) When the Cousteau team celebrated on *Conshelf II* at 12 m below the sea surface in 1963, they used Champagne (sparkling wine). Unfortunately, it tasted flat as the high pressure in the cabin did not allow the bubbles to escape the wine.

(III) There are also some more serious dangers of heliox. For example, the heat loss of the human body is increased by a factor of 70 in comparison to air.

This question shows how a topic of interest in one context rapidly branches out into other areas of the physical sciences.

Question 9.5

(a) The osmolarity of a solution is measured with an osmometer. Does Question 9.4 provide a physical principle suitable for designing an osmometer?

(b) The osmolarity in an aqueous solution is recorded as a value per litre water. If you measure osmolarity for blood plasma that contains 70 g/L proteins, or for cytoplasm in erythrocytes with 300 g/L haemoglobin, how large a correction factor do you need to include in the value you report?

Answer to part (a): An osmometer is a device that measures the lowering of the freezing point of a solution. The concentration of osmotically active components in the solution is proportional to ΔT in Fig. 9.19(b).

Answer to part (b): The osmometer measures osmolarity in unit osm/kg of water, not osm/L of water. These two values are identical for dilute solutions; however, they have to be corrected for real solutions. For example, 1 L of water is 1.0 kg, but 1 kg of blood plasma contains only 0.93 kg water, with the remaining 0.07 kg = 70 g present in the form of proteins. This effect is even larger within a cell, with 1 L intracellular fluid containing only 0.7 kg water (and 0.3 kg haemoglobin in the case of erythrocytes).

ANALYTICAL PROBLEMS

Problem 9.1

The osmotic effect is often used to determine the molar mass of macromolecules. Here we illustrate how this is done. The apparatus used is shown in Fig. 9.18: it consists of two chambers that are separated by a semipermeable membrane. One tube is filled with a dilute solution of the macromolecules and the other tube is filled with pure solvent. Additional tubes are mounted vertically on each chamber to measure the osmotic pressure (using Pascal's law, as discussed in Chapter 11).

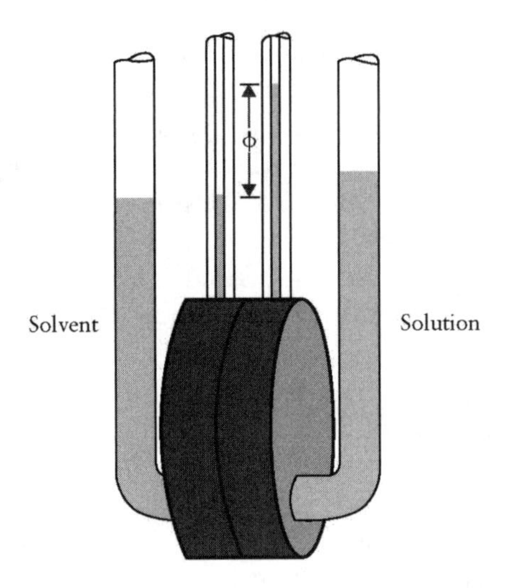

FIGURE 9.18

Assume that the experiment is done at 25°C. We use van't Hoff's law in the form:

$$\Pi = \frac{R \cdot T}{M} c \quad \textbf{(2)}$$

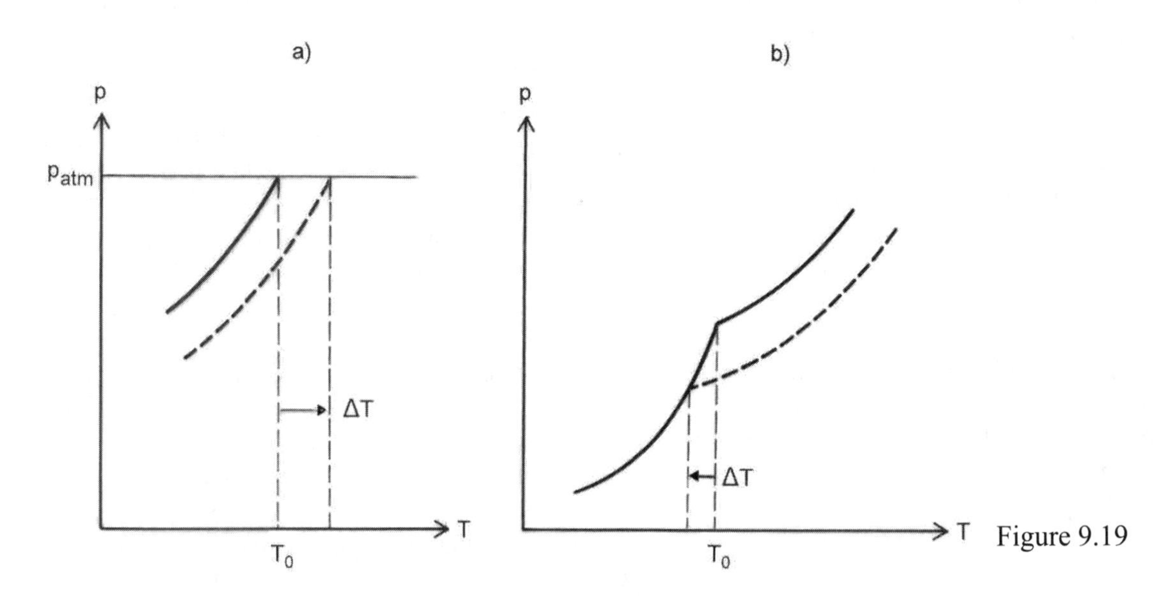

Figure 9.19

Table 9.1

concentration (g/cm³)	osmotic pressure Π (atm)	
	in benzene	in cyclohexane
0.02	0.0021	0.0122
0.015	0.00153	0.0068
0.01	0.001	0.0031
0.005	0.0005	0.0009

where c is the concentration of the macromolecule in the solution (in unit kg/m³), M is the molar mass (in unit kg/mol) of the macromolecule and Π is the osmotic pressure, as measured with the apparatus in Fig. 9.18. Since van't Hoff's law applies exactly only for very dilute solutions, several measurements are taken for various dilute solutions (i.e., several different values of c) and are then extrapolated. For the extrapolation, we plot Π/c vs. c and obtain the value for Π/c at $c = 0$ from the plot.

Using the data from Table 9.5 for the osmotic pressure of polyisobutylene in benzene and cyclohexane
(a) plot Π/c versus c for both solutions,
(b) find the extrapolation value of Π/c at $c = 0$ for each curve, and
(c) use Eq. [2] to determine the molar mass of polyisobutylene.

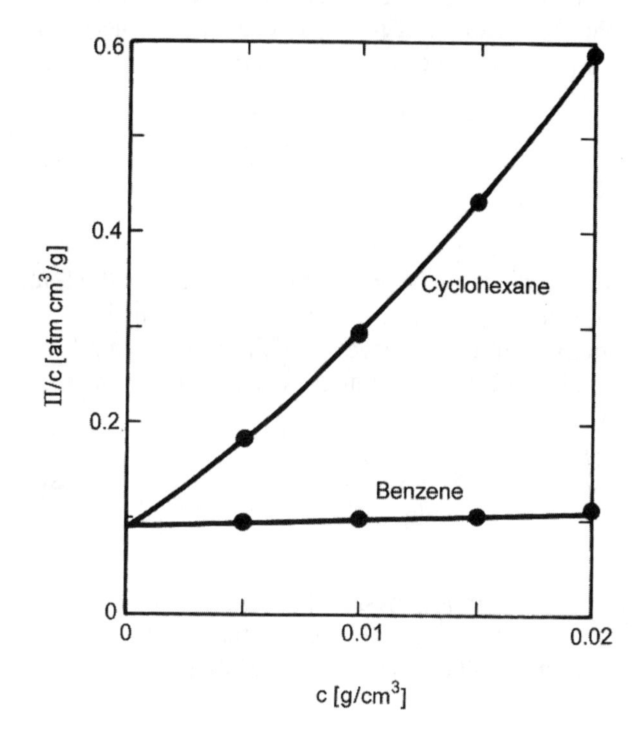

Figure 9.20

Solution to part (a): The best way to prepare the plot for Π/c vs. c is to add a column to Table 9.5 to provide the term Π/c. This is done in Table 9.6. With the data in Table 9.6, the plot in Fig. 9.20 is obtained.

Table 9.6

c (g/cm³)	Π/c (atm · cm³/g)	
	benzene	cyclohexane
0.0200	0.105	0.61
0.0150	0.102	0.453
0.0100	0.1	0.31
0.0050	0.1	0.18

Solution to part (b): The plot illustrates that both curves indeed converge when extrapolated to $c = 0$ g/cm³. This establishes the validity of the current approach. We read the extrapolated value for Π/c from Fig. 9.20 (alternatively, numerical extrapolation methods may be used by entering the data of Table 9.6 in a computer):

$$\lim_{c \to 0} \frac{\Pi}{c} = 0.097 \frac{atm \cdot cm^3}{g} \quad \textbf{(3)}$$

Now we convert Eq. [3] into standard units:

$$\left(0.097 \frac{atm \cdot cm^3}{g}\right)\left(1.013 \times 10^5 \frac{Pa}{atm}\right) \times \left(10^{-6} \frac{m^3}{cm^3}\right)\left(10^3 \frac{g}{kg}\right) \quad \textbf{(4)}$$

Thus, Eq. [3] becomes:

$$\lim_{c \to 0} \frac{\Pi}{c} = 9.83 \frac{Pa \cdot m^3}{kg} \quad \textbf{(5)}$$

Solution to part (c): We first confirm that the relation given for van't Hoff's law in this problem is indeed equivalent to Eq. [1]. For this we rewrite the mole fraction in Eq. [1] using Eq. [9.6]:

$$\Pi = \frac{RT}{V_1^0} \frac{n_2}{n_{total}} = RT \frac{n_2}{V_1^0} \quad \textbf{(6)}$$

where the second equation results because the total amount of liquid in the system is $n_{total} = 1.0$ mol since van't Hoff's law is derived for one mol of solution. We replace the amount of the dissolved macromolecule in Eq. [6] with its concentration, using:

$$c_2 = \frac{n_2 \cdot M_2}{V} \cong \frac{n_2 \cdot M_2}{V_1^0} \quad \textbf{(7)}$$

in which we assume for the last equation that the total volume and the volume of solvent component 1 are essentially identical since we are dealing with a very dilute solution. Now we substitute Eq. [7] into Eq. [6] and find:

$$\Pi = RT \frac{n_2}{V_1^0} = RT \frac{c_2}{M_2} \quad \textbf{(8)}$$

To obtain the molar mass of the macromolecule we rewrite Eq. [8], taking into account that the formula applies in the limit of a very dilute solution:

$$M_2 = R \cdot T \left[\lim_{c_2 \to 0} \left(\frac{\Pi}{c_2} \right) \right]^{-1}$$

$$= \frac{\left(8.314 \frac{J}{K \cdot mol} \right) (298\ K)}{9.83 \frac{Pa \cdot m^3}{kg}} = 252 \frac{kg}{mol} \quad \textbf{(9)}$$

The result is usually expressed in unit g/mol as $M_2 = 252000$ g/mol.

Problem 9.3

Use Raoult's law to derive the vapour pressure depression for an ideal dilute solution. *Hint*: consider a system with two components, solvent A and solute B with $x_A >> x_B$.

Solution: Write $x_B = 1 - x_A$ in Raoult's law:

$$\frac{p_A}{p_A^0} = 1 - x_B \quad \textbf{(10)}$$

which yields:

$$x_B = \frac{p_A - p_A^0}{p_A^0} = \frac{\Delta p}{p_A^0} \quad \textbf{(11)}$$

i.e., $\Delta p \propto x_B$. The proportionality factor is the (inverse) vapour pressure of the pure solvent.

CHAPTER TEN

Membranes: Transport of Energy and Matter

MULTIPLE CHOICE AND CONCEPTUAL QUESTIONS

Question 10.1

Which of the following statements is true? (A) Gold conducts heat better than copper. (B) Under otherwise equal conditions, about 2 times more heat transfers through gold than through iron per second. (C) If I conduct Fourier's experiment with a steel and an aluminium rod of same shape, then I must use a longer rod for aluminium to obtain the same rate of heat transfer between beakers of freezing and boiling water. (D) The thermal conductivity in Table 10.1 is not given in SI units. (E) Table 10.1 shows that the thermal conductivity is a materials–independent constant.

Answer: (C)

Table 10.1

Material	Thermal Conductivity λ ($J\ s^{-1}\ m^{-1}\ K^{-1}$)
1. Solid metals and alloys	
Silver (Ag)	420
Copper (Cu)	390
Gold (Au)	310
Aluminium (Al)	230
Iron (Fe)	80
Steel	50
2. Nonmetallic solids	
Ice	1.6
Quartz glass (SiO_2)	1.4
Window glass	0.8
Fat	0.24
Rubber	0.2
Wood	0.12 – 0.04
Felt, Silk	0.04
3. Liquids	
Mercury (Hg)	8.3
Water (H_2O)	0.6
Ethanol (C_2H_5OH)	0.18
4. Gases	
Air	0.026

Question 10.3

The material of the rod in Fourier's experiment is changed such that its thermal conductivity decreases by 20%. What change allows us to best reestablish the previous flow rate of heat (Q/t): (A) decreasing the length of the rod by 10%, (B) increasing the length of the rod by 10%, (C) increasing the diameter of the rod by more than 20%, (D) increasing the diameter of the rod by 20%, (E) increasing the diameter of the rod by less than 20%.

Answer: (E) The flow rate of heat is proportional to the cross–sectional area of the rod, and so is proportional to the square of the diameter.

Question 10.5

When studying the heat loss of a sphere of radius R we use the following term for the surface area A through which heat flows in Fourier's law: (A) $A = \pi \cdot R$, (B) $A = \pi \cdot R^2$, (C) $A = 4 \cdot \pi \cdot R^2$, (D) $A = 4 \cdot \pi \cdot R^3$, (E) $A = 4 \cdot \pi \cdot R^3/3$.

Answer: (C)

Question 10.7

The geothermal effect states that the temperature below Earth's surface increases by one degree Celsius for every 30 metres depth. Assuming a surface temperature of 0^0C, which of the following statements is true: The temperature rises to 1000 K at (choose the closest value) (A) 8 km depth, (B) 15 km depth, (C) 22 km depth, (D) 30 km depth, (E) 100 km depth.

Answer: (C). $(1000\ K - 273\ K)\ (30\ m/K) = 22 \times 10^3$ m.

Question 10.9

The column of mercury in a thermometer initially descends slightly before rising when the instrument is placed in a hot liquid. Why?

Answer: The mercury is contained in a glass cylinder. Heat flows through the glass cylinder to the metal; the temperature rise of the glass occurs first, leading to its expansion. This causes an increase in the open volume inside and the mercury settles down (like any other liquid based on its mechanical equilibrium, see Chapter 11).

Question 10.11

Use the diffusion coefficients for sucrose and tobacco mosaic virus in water from Table 10.5. To obtain the same rate of mass transfer for equal concentrations of sucrose and the virus on both sides of a water–filled cylindrical tube of fixed tube radius, the tube length has to be reduced for the virus by a factor of (choose the closest value): (A) 3.2, (B) 10, (C) 32, (D) 100, (E) 320.

Answer: (D)

Table 10.5

System	Diffusion coefficient
Oxygen (O_2) in air	6.4×10^{-5} m²/s
Oxygen (O_2) in water	1×10^{-9} m²/s
Oxygen (O_2) in tissue	1×10^{-11} m²/s
Water in water	2.4×10^{-9} m²/s
Sucrose in water	5×10^{-10} m²/s
Haemoglobin in water	7×10^{-11} m²/s
Tobacco mosaic virus in water	5×10^{-12} m²/s

Question 10.13

In two separate experiments the following two diffusion coefficients of a contaminant in tissue are found: experiment I: $D = 7 \times 10^{-11}$ m²/s and experiment II: $D = 7 \times 10^{-9}$ cm²/s. Your laboratory head suggests further experiments to check whether the two contaminants are the same. What do you do? (A) Conduct the suggested experiments because the two diffusion coefficients are essentially the same. (B) Repeat the previous experiments to see whether the new data are still so close to each other. (C) Ask another researcher to confirm the group head's conclusion because it isn't that easy to compare the two given diffusion coefficients. (D) Reject the laboratory head's suggestion and proceed with the conclusion that the two contaminants are different.

Answer: (D). If the measurements you took are both accurate and precise, there is no doubt that there is a significant difference between the two diffusion coefficients. Though this author would likely ask the head if the purpose of running the experiment again was to check that the measurements were both accurate and precise.

Question 10.15

The diffusion length of a molecule in a solution is Λ = 1 cm after the experiment is run for t = 1 hour. When will the diffusion length double? (A) after an additional hour, (B) after an additional 2 hours, (C) after an additional 3 hours, (D) after an additional 4 hours.

Answer: (C). It takes four times as long to go twice the diffusion length. That means it will take four hours in total; since we have already waited one hour, we have an additional 3 hours to wait.

Question 10.17

Which of the following statements about a system in steady state is correct? (A) The system is in an equilibrium state. (B) All system parameters change as a function of time. (C) The internal energy of the system increases continuously. (D) The entropy production in the system is a minimum. (E) The system reaches the equilibrium within a short time.

Answer: (D)

ANALYTICAL PROBLEMS

Problem 10.1

We quantify Fourier's experiment, shown in Fig. 10.4, for a cylindrical copper rod of a length 1.2 m and a cross–sectional area 4.8 cm². The rod is insulated to

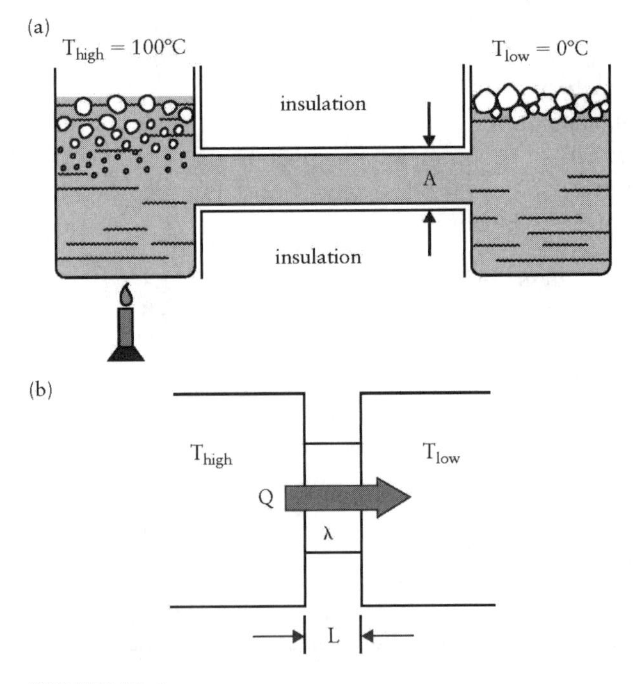

FIGURE 10.4

prevent heat loss through its surface. A temperature difference of 100 K is maintained between the ends. Find the rate at which heat is conducted through the rod.

Solution: Fourier's law is used with the thermal conductivity for copper taken from Table 10.1:

$$\frac{Q}{t} = \lambda_{Cu} \cdot A \frac{\Delta T}{l}$$

$$= \frac{\left(390 \frac{J}{m\,s\,K}\right)(4.8 \times 10^{-4}\,m^2)(100\,K)}{1.2\,m} \quad \textbf{(1)}$$

$$= 15.6 \frac{J}{s}$$

Note that this is less than a tenth of the amount of heat that your body loses in the winter with dry clothing (see Example 10.1). Why would we compare these results? The answer is that we build an intuition for such numbers, as the loss of heat in winter is a process to which we can relate while the heat flowing in a metal rod is elusive.

Problem 10.3

For poor heat conductors the thermal resistance R has been introduced. The thermal resistance of a piece of material of thermal conductivity λ and thickness l is defined as:

$$R = \frac{l}{\lambda} \quad \textbf{(2)}$$

(a) Show that Eq. [2] allows us to rewrite Fourier's law in the form:

$$\frac{Q}{t} = A \frac{T_{high} - T_{low}}{R} \quad \textbf{(3)}$$

in which A is the cross–sectional area of the piece of material.

(b) What is the SI unit of the thermal resistance R?

Solution to part (a): Eq. [3] is obtained by substituting Eq. [2] in the form $\lambda/L = 1/R$ into Fourier's law:

$$\frac{Q}{t} = \lambda \cdot A \frac{T_{high} - T_{low}}{l} \quad \textbf{(4)}$$

Solution to part (b): We determine the unit of R from Eq. [2]:

$$unit\,(R) = \frac{unit\,(l)}{unit\,(\lambda)}$$

$$= \frac{m}{\frac{J}{m \cdot s \cdot K}} = \frac{s \cdot K \cdot m^2}{J} \quad \textbf{(5)}$$

Problem 10.5

In a cookbook, a poultry thawing chart states that a 10-kg whole turkey takes four days to defrost in a refrigerator. Estimate how long it would take to defrost a 2-tonne Siberian mammoth from the same initial temperature in an industrial refrigeration hall. *Hint*: treat both animals as spherically shaped and use the same approach we applied in Example 10.3.

Solution: We isolate the time in Eq. [4]:

$$t = \frac{Q}{\lambda \cdot A \frac{\Delta T}{l}} \quad \textbf{(6)}$$

i.e., the time for the process is proportional to the amount of heat required and inversely proportional to the temperature gradient in the meat. The required heat is written with Joule's definition $Q = c \cdot m \cdot \Delta T$:

$$t = \frac{c \cdot m \cdot \Delta T}{\lambda \cdot A \frac{\Delta T}{l}} \quad \textbf{(7)}$$

The specific heat capacity and the thermal conductivity of turkey and mammoth are assumed to be the same. To eliminate these terms, we rewrite Eq. [7] as a ratio:

$$\frac{t_{mammoth}}{t_{turkey}} = \frac{\frac{m_{mammoth}}{\left(\frac{A}{l}\right)_{mammoth}}}{\frac{m_{turkey}}{\left(\frac{A}{l}\right)_{turkey}}} \quad \textbf{(8)}$$

The mass scales with L^3 where L is the size (typical length) for an object of uniform density. This leads to $A/l \propto L \propto m^{1/3}$. Thus, we rewrite Eq. [8] in terms of the respective body masses:

$$\frac{t_{mammoth}}{t_{turkey}} = \frac{m_{mammoth}^{2/3}}{m_{turkey}^{2/3}} = \left(\frac{2000\ kg}{10\ kg}\right)^{2/3} = 34 \quad \textbf{(9)}$$

Defrosting a mammoth takes 136 days.

Problem 10.7

Fig. 10.32 shows a block that consists of two materials with different thicknesses l_1 and l_2 and different thermal conductivities λ_1 and λ_2. The temperatures of the outer surfaces of the block are T_{high} and T_{low}, as shown in the figure. Each face of the block has a cross–sectional area A.

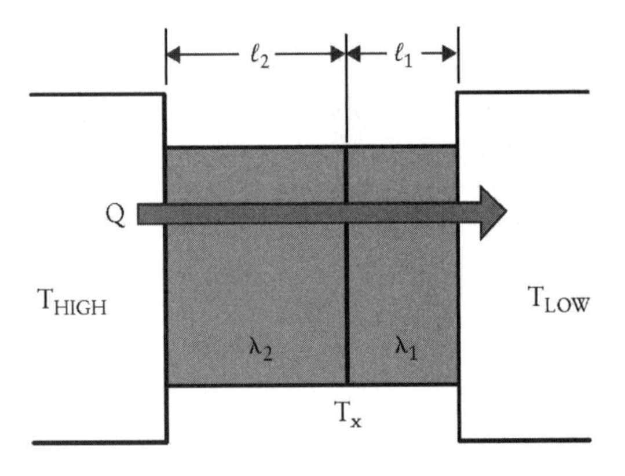

FIGURE 10.32

(a) Show that the formula

$$\frac{Q}{t} = \frac{A\ (T_{high} - T_{low})}{(l_1/\lambda_1) + (l_2/\lambda_2)} \quad \textbf{(10)}$$

correctly expresses the steady–state rate of heat transfer. *Hint*: in the steady state the heat transfer through any part of the block must be equal to the heat transfer through the other part of the block. Introduce a temperature T_x at the interface of the two parts as shown in Fig. 10.32, and then express the rate of heat transfer for each part of the block separately.

(b) Rewrite Eq. [10] using Eq. [2], which introduces R_1 and R_2 as the thermal resistances for the two parts of the block. By comparing the result with Eq. [3], determine how thermal resistances are combined for materials in sequence.

Solution to part (a): In this problem, the underlying assumption is that the heat transfer is a steady process, i.e., that (i) the temperature profile across the block and (ii) the rate of heat transfer do not change with time. Later in the textbook, we will define this as a distinguished state of the system called the *steady state*. In the steady state, the heat transferred through the two blocks per unit time must be the same. Otherwise, we would either have a depletion of heat at the interface between both blocks and the interface zone would have to cool down or we would have an accumulation of heat at the interface and that zone would become hotter and hotter.

As a consequence, we can use Fourier's law twice, once for each block. The implication of the steady state requires $(Q/t)_1 = (Q/t)_2 = Q/t$, in which the indices 1 and 2 represent each of the blocks. Labelling the temperature at the interface T_x we obtain:

$$\frac{Q}{t} = \frac{\lambda_2 A \left(T_{high} - T_x\right)}{l_2} = \frac{\lambda_1 A \left(T_x - T_{low}\right)}{l_1} \quad \textbf{(11)}$$

Note that Eq. [11] contains three formulas, as the first and second or first and third or second and third terms are equal. First, we use the equality of the second and third terms to derive a formula for the temperature at the interface, T_x. Multiplying both sides of the equation with $l_1 \cdot l_2$ and dividing both sides by the area A, we find:

$$\lambda_2 l_1 T_{high} - \lambda_2 l_1 T_x = \lambda_1 l_2 T_x - \lambda_1 l_2 T_{low} \quad \textbf{(12)}$$

Next we group together the terms containing T_x:

$$T_x\left(\lambda_1 l_2 + \lambda_2 l_1\right) = \lambda_2 l_1 T_{high} + \lambda_1 l_2 T_{low} \quad \textbf{(13)}$$

In a last step we isolate T_x:

$$T_x = \frac{\lambda_2 l_1 T_{high} + \lambda_1 l_2 T_{low}}{\lambda_1 l_2 + \lambda_2 l_1} \quad \textbf{(14)}$$

This result is substituted into either one of the other two formulas in Eq. [11]. Choosing the equality between the first and the second terms, the following calculation follows. For convenience, we do not substitute Eq. [14] into that equation right away since the formula would become very long and cumbersome. Instead, we initially substitute Eq. [14] only into the term $(T_{high} - T_x)$, which is a factor in the second term in Eq. [11]. Once this bracket has been simplified, we return to the full formula in Eq. [11] and determine the heat flow rate Q/t. For the bracket $(T_{high} - T_x)$ we find:

$$T_{high} - T_x = T_{high} - \frac{\lambda_2 l_1 T_{high} + \lambda_1 l_2 T_{low}}{\lambda_1 l_2 + \lambda_2 l_1}$$

$$= \frac{\lambda_1 l_2 T_{high} + \lambda_2 l_1 T_{high} - \lambda_2 l_1 T_{high} - \lambda_1 l_2 T_{low}}{\lambda_1 l_2 + \lambda_2 l_1} \quad \textbf{(15)}$$

$$= \frac{\lambda_1 l_2 (T_{high} - T_{low})}{\lambda_1 l_2 + \lambda_2 l_1}$$

The result in Eq. [15] is now substituted for the term $(T_{\text{high}} - T_x)$ in the first equality of Eq. [11]:

$$\frac{Q}{t} = \frac{\lambda_2 A}{l_2} \frac{\lambda_1 l_2 (T_{high} - T_{low})}{\lambda_1 l_2 + \lambda_2 l_1}$$

$$= A \frac{\lambda_1 \lambda_2}{\lambda_1 l_2 + \lambda_2 l_1} (T_{high} - T_{low}) \quad \textbf{(16)}$$

Lastly, we divide both the numerator and the denominator of the right hand side of Eq. [16] by $\lambda_1 \cdot \lambda_2$, leading to:

$$\frac{Q}{t} = A \frac{(T_{high} - T_{low})}{\frac{l_2}{\lambda_2} + \frac{l_1}{\lambda_1}} \quad \textbf{(17)}$$

which is the result we sought.

Solution to part (b): Eq. [17] is rewritten by applying Eq. [2] to each of the blocks, i.e., $R_1 = l_1/\lambda_1$ and $R_2 = l_2/\lambda_2$:

$$\frac{Q}{t} = A \frac{(T_{high} - T_{low})}{R_1 + R_2} \quad \textbf{(18)}$$

Thus, the R values, which are the thermal resistances, are added for blocks through which heat is passing in sequence. When we study fluid flow through tubes placed in sequence (Chapter 12), we will find that fluid flow resistances are also added; in Chapter 14 we will find the same to apply to electric resistances, which are added when electric charges flow through resistors in series. This illustrates that our finding is broadly applicable: resistances have to be added if the flow is through several components in sequence. We will also find that flow resistances have to be added inversely when the components are placed in parallel, i.e., such that the flow is divided and its parts pass through the components simultaneously.

Problem 10.9

We want to compare combinations of insulator materials in the case shown in Fig. 10.33. The figure shows a two–layer system with a total of four square pieces. Each piece has an area A, thus the two systems each cover an area of $2 \cdot A$. The two arrangements in Fig. 10.33 differ in the order of the two materials (labelled 1 and 2). Which arrangement, (a) or (b), allows for a greater heat flow?

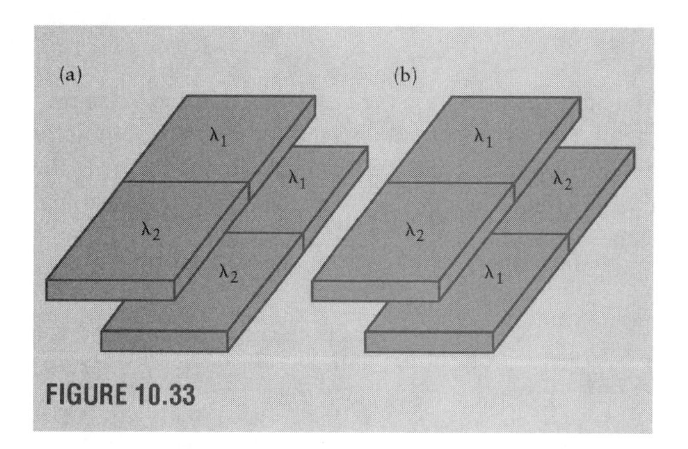

FIGURE 10.33

Solution: We start with arrangement (a). After combining the two blocks we obtain an insulation layer of area $2 \cdot A$ with 50% of this area covered with material 1 of (combined) thickness $2 \cdot l$ where l is the thickness of a single piece of insulator. For this arrangement Fourier's law reads:

$$\left.\frac{Q}{t}\right|_{(a)} = \left(\frac{Q}{t}\right)_1 + \left(\frac{Q}{t}\right)_2$$

$$= \frac{A \lambda_1}{2 l} (T_{high} - T_{low}) + \frac{A \lambda_2}{2 l} (T_{high} - T_{low}) \quad \textbf{(19)}$$

We rewrite Eq. [19] for later comparison in the form:

$$\left.\frac{Q}{t}\right|_{(a)} = \frac{A}{2 l} \lambda_{comb}^{(a)} (T_{high} - T_{low})$$

$$with: \quad \lambda_{comb}^{(a)} = \lambda_1 + \lambda_2 \quad \textbf{(20)}$$

Now we consider arrangement (b). Both half–sides of the combined insulator consist of two layers with a 1–2 sequence of the two materials. To describe such an arrangement, we derived Eq. [10]. Using that formula for a total area of $2 \cdot A$ in the current problem, we write:

$$\frac{Q}{t}\Big|_{(b)} = \frac{2 \cdot A\left(T_{high} - T_{low}\right)}{\frac{l}{\lambda_1} + \frac{l}{\lambda_2}} \quad \textbf{(21)}$$

which we rewrite in analogy to Eq. [20] above:

$$\frac{Q}{t}\Big|_{(b)} = \frac{A}{2\,l}\lambda^{(b)}_{comb}\left(T_{high} - T_{low}\right)$$
$$with: \quad \lambda^{(b)}_{comb} = \frac{4 \cdot \lambda_1 \cdot \lambda_2}{\lambda_1 + \lambda_2} \quad \textbf{(22)}$$

in which the factor 4 in the last formula is due to the pre-factor 2 moving from the numerator to the denominator between Eqs. [21] and [22]. The remaining factors in the formula for $\lambda^{(b)}{}_{comb}$ are due to the following mathematical step:

$$\frac{1}{\frac{1}{\lambda_1} + \frac{1}{\lambda_2}} = \frac{1}{\frac{\lambda_1 + \lambda_2}{\lambda_1 \cdot \lambda_2}} = \frac{\lambda_1 \cdot \lambda_2}{\lambda_1 + \lambda_2} \quad \textbf{(23)}$$

In order to compare Eqs. [20] and [22], we divide Eq. [20] by Eq. [22]. Since all other terms in both equations are equal, this corresponds to a division of the two combined λ factors:

$$\frac{\lambda^{(a)}_{comb}}{\lambda^{(b)}_{comb}} = \frac{\left(\lambda_1 + \lambda_2\right)^2}{4 \cdot \lambda_1 \cdot \lambda_2} \quad \textbf{(24)}$$

We use a mathematical trick to determine whether the ratio in Eq. [24] is larger or smaller than 1: since we know that the square of any number is positive the following holds:

$$\begin{aligned}\left(\lambda_1 - \lambda_2\right)^2 &= \lambda_1^2 - 2 \cdot \lambda_1 \cdot \lambda_2 + \lambda_2^2 \\ &= \lambda_1^2 + 2\,\lambda_1\,\lambda_2 + \lambda_2^2 - 4\,\lambda_1\,\lambda_2 \\ &= \left(\lambda_1 + \lambda_2\right)^2 - 4 \cdot \lambda_1 \cdot \lambda_2 > 0\end{aligned} \quad \textbf{(25)}$$

which is true for any pair of values λ_1 and λ_2 if we assume $\lambda_1 \neq \lambda_2$. Note that the case $\lambda_1 = \lambda_2$ is trivial and not discussed further. Therefore, we find the following inequality: $(\lambda_1 + \lambda_2)^2 > 4 \cdot \lambda_1 \cdot \lambda_2$. From this we get:

$$\frac{\lambda^{(a)}_{comb}}{\lambda^{(b)}_{comb}} > 1 \quad \Rightarrow \quad \lambda^{(a)}_{comb} > \lambda^{(b)}_{comb} \quad \textbf{(26)}$$

The larger combined thermal conductivity for arrangement (a) leads to a larger amount of heat transported per time unit.

Problem 10.11
In the text we discussed the Australian Brush–turkey's approach to nesting. We noted that it must maintain the radius of a decomposing forest–litter mound in which it eggs incubate according to Eq. [10.12]:

$$R \propto \sqrt{T_{incubation} - T_{ambient}} \quad \textbf{(27)}$$

Confirm that this equation properly describes the relation between mound radius and its temperature profile.

Solution: We assume that the bacterial putrefaction generates heat at a uniform rate, $(Q_{putrefaction}/t)/V = const = q^*$. Thus, the total amount of heat generated per second in a mound of radius R is:

$$\frac{Q_{putrefaction}}{t} = \frac{4}{3}\pi \cdot R^3 \cdot q^* \quad \textbf{(28)}$$

In the stationary case sought by the birds, this rate must be equal to the heat loss through the mound surface. Assuming that heat loss occurs uniformly across the surface of the sphere, we write:

$$\frac{4}{3}\pi \cdot R^3 \cdot q^* = -4 \cdot \pi \cdot R^2 \cdot \lambda \frac{\Delta T}{\Delta R} \quad \textbf{(29)}$$

where the last term is Fourier's heat loss rate in which we rewrote the temperature step across the membrane of thickness l, $\Delta T/l$, as a temperature gradient $\Delta T/\Delta R$. From Eq. [29] we find the required temperature gradient:

$$\frac{\Delta T}{\Delta R} = -\frac{R \cdot q^*}{3 \cdot \lambda} \quad \textbf{(30)}$$

If you want to avoid the use of calculus, you use a similar line of arguments from Eq. [30] as we use in the textbook on page 510 to derive the elastic energy from the linear restoring force for an object attached to a spring. We find that a gradient that is linear in the position, $\Delta T/\Delta R \propto R$, yields a quadratic formula for the temperature profile with radius:

$$T - T_0 = -\frac{R^2 \cdot q^*}{6 \cdot \lambda} \quad \textbf{(31)}$$

The particular application for the Australian Brush–turkey reads as follows: the central temperature T_0 must be maintained at 35°C, but the external temperature T may vary. To compensate for variations of the external air temperature, the birds have only control over the radius of the mound; they must adjust the radius based on:

$$R = \sqrt{\frac{6 \cdot \lambda}{q^*}(T_{eggs} - T_{air})} \quad \textbf{(32)}$$

which is consistent with Eq. [27] because λ and q^* are constant.

Problem 10.13

Heat loss via convection occurs only when heat is carried by a moving fluid. For example, when heating water in a beaker from below, the increase of the water temperature at the bottom leads to a decrease of the water density and causes the warmer water to rise due to buoyancy. The rising water carries excess heat to the surface.

(a) Compare bare skin to skin covered with clothes. Why is the heat loss of the body significantly reduced when wearing clothes?

(b) At temperate lakes and ponds it is often observed that algae bloom for a short period during spring and autumn. Consider Fig. 10.35 which shows the stratification during summer (top) as well as the seasonal turnover in spring and autumn (bottom), how can the convection–driven turnover cause algal blooms?

Solution to part (a): A major heat loss mechanism at moderate environmental temperatures is convection. It requires that an air flow pass across the skin to carry the heat away. Clothes have air–filled pores of variable size in which the air cannot move. Thus, heat which would otherwise be lost from the bare skin by convection, must be conducted through air in the pores of the clothes. Heat conduction (Fourier's law) is a much less effective mechanism than convection, particularly since the thermal conductivity coefficient λ of air is very small.

Solution to part (b): The algal blooms that one can observe in temperate lakes during the transitional seasons are caused by convective currents in the water of the lake. As illustrated in Fig. 10.35, the temperature profile of the lake during summer is layered with (i) a warm zone near the surface (since this zone is heated by the warm air and the sun's radiation), (ii) an intermediate layer called *thermocline*, and (iii) a cold layer near the bottom of the lake. During the transitional seasons the temperature change in the upper layer causes the thermocline layer to break down, which in turn causes convection to mix the water. The water currents associated with convection transport nutrients from the bottom of the lake to the surface. The nutrient–rich water brought that way to the surface can cause a rapid increase in the algae population as algae also need sunlight to grow.

Problem 10.15

Determine the diffusion coefficient for glycerine in H_2O using the following observations: glycerine diffuses along a horizontal, water–filled column that has a cross–sectional area of 2.0 cm². The density step across $\Delta\rho/l$ is 3.0×10^{-2} kg/m⁴ and the steady–state diffusion rate is 5.7×10^{-15} kg/s.

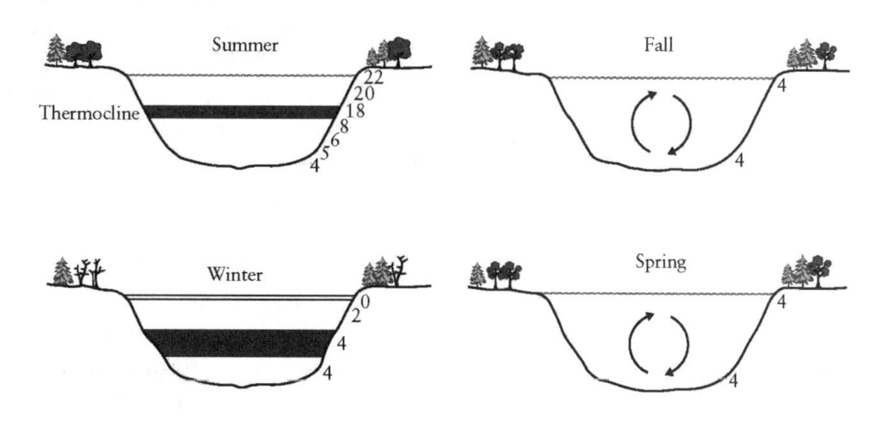

FIGURE 10.35

Solution: We rewrite Fick's law with the diffusion coefficient as the dependent variable:

$$D = \frac{1}{A}\left(\frac{m}{t}\right)\frac{1}{\Delta\rho/l} \qquad \textbf{(33)}$$

where $\Delta\rho/l$ is the change of the density per unit length, i.e., the density gradient. In the last step, the given data are substituted into Eq. [33]:

$$D = \frac{5.7 \times 10^{-15}\,\frac{kg}{s}}{(2 \times 10^{-4}\,m^2)\left(3 \times 10^{-2}\,\frac{kg}{m^4}\right)} = 9.5 \times 10^{-10}\,\frac{m^2}{s} \qquad \textbf{(34)}$$

Problem 10.17

Why can bacteria rely on passive diffusion for their oxygen supply but human beings cannot? *Hint*: Calculate from Eq. [10.34]:

$$\Lambda = \sqrt{2 \cdot D \cdot t} \qquad \textbf{(35)}$$

(a) the time it takes for oxygen to diffuse from the interface with the environment to the centre of a bacterium of radius $r = 1.0\ \mu m$, and

(b) the time it takes for oxygen to diffuse from the external air to an organ 10 cm below human skin. *Note*: for an upper limit use the diffusion coefficient of oxygen in water, and for a lower limit use the diffusion coefficient of oxygen in tissue from Table 10.5. These two values give you a good approximation since humans consist of roughly 10 L extracellular fluid and 30 L cells.

(c) *If you are interested*: Why can many relatively large invertebrates such as hydras survive without a cardiovascular system?

Solution to part (a): Aerobic bacteria need to have oxygen present throughout their body in order for their metabolisms to function. Let us assume a spherical bacterium of average size, e.g. with a radius of 1 μm (although they are often not spherical, as their name implies). For this bacterium we calculate the time that it takes for oxygen to diffuse from the outer surface to the centre, using Eq. [35]. For the diffusion length we use $\Lambda = 1.0 \times 10^{-6}$ m. We obtain a lower limit for the diffusion time by using the diffusion coefficient for oxygen in water at room temperature from Table 10.5:

$$t = \frac{\Lambda^2}{2\,D} = \frac{(1.0 \times 10^{-6}\,m)^2}{2\left(1.0 \times 10^{-9}\,\frac{m^2}{s}\right)} = 5 \times 10^{-4}\,s \qquad \textbf{(36)}$$

which is just half a millisecond.

In the same way, we use the diffusion coefficient of oxygen in tissue at room temperature from Table 10.5 and obtain an upper limit of $t = 5 \times 10^{-2}$ s, which is still less than a tenth of a second. Thus, an average sized bacterium at room temperature has no problem assimilating the oxygen it needs for its metabolism through passive diffusion.

Solution to part (b): Now we focus on the human body with an oxygen consuming organ, such as the heart, at an assumed depth of 10 cm below the surface of the outer skin. This leads to a required diffusion length $\Lambda = 0.1$ m for this organ to be supplied with oxygen. We calculate the same two limiting cases as for the bacteria, assuming that oxygen is brought to the organ by passive diffusion. Note that a body temperature of endotherms of 37^0C instead of 20^0C improves the following result only insignificantly. For diffusion in water we find:

$$t = \frac{\Lambda^2}{2\,D} = \frac{(0.1\,m)^2}{2\left(1.0 \times 10^{-9}\,\frac{m^2}{s}\right)} = 5 \times 10^{6}\,s \cong 58\ days \qquad \textbf{(37)}$$

and for diffusion in tissue we obtain $t = 5 \times 10^8$ s, which is just less than 16 years. This result shows clearly that a larger organism cannot rely on passive diffusion to provide the oxygen required for its metabolism.

It is interesting to put this result in context with Table 10.4, which provides the history of life on Earth. When you read that table carefully, you notice that it took just over 500 million years for life to actually emerge after conditions that permit life to exist developed. It took five times this long, a staggering 2.5 billion years, to have the first single–celled bacteria develop into a larger, multi–celled organisms. Once this had been achieved, a rapid diversification of life took place about 670 million years ago, called the Cambrian Explosion. Thus, Einstein's equation for the diffusion length can be considered to represent the single biggest hurdle in the development of life, not the occurrence of life itself. It is noteworthy that material transport within cells remains diffusion–limited, and thus, even the largest mammalian cells have diameters of no more than 20 μm.

Table 10.4

Years ago	Event
13×10^9	Age of universe (Big Bang)
4.7×10^9	Formation of the solar system from an interstellar cloud of gas
4.6×10^9	Proto–Earth (great bombardment)
4.03×10^9	Oldest rock (Yellowknife, Canada)
$3.6–3.8 \times 10^9$	First prokaryote (stromatolithic bacteria)
2.5×10^9	First eukaryote (algae)
1.7×10^9	Oxygen atmosphere
1.0×10^9	Sexual reproduction
6.7×10^8	Multi–celled animal fossils found at many places on Earth
5.8×10^8	Animals with shells and skeleton
4.8×10^8	Plants expand from sea to land
4.2×10^8	Animals expand from sea to land
2.4×10^8	First mammals
65×10^6	End of dinosaurs
4.0×10^6	Early hominids (Australopithecus)
2.5×10^6	Genus homo
0.125×10^6	Modern homo sapiens

Solution to part (c): Hydras possess a central gastro–vascular cavity that is lined by a layer that is two cells thick to allow diffusive material exchange.

Problem 10.19

(a) How far does a tobacco mosaic virus move in water at 20⁰C in 1 hour?
(b) Using the ratio of the diffusion coefficients for oxygen and carbon dioxide in tissue from Problem 10.16, what is the ratio of diffusion lengths for these molecules in tissue at 20⁰C?

Solution to part (a): We use Eq. [35]. The diffusion coefficient of the tobacco mosaic virus in water is taken from Table 10.9:

Table 10.9

Protein	D (m²/s)	M (kg/mol)
Tobacco mosaic virus	5.3×10^{-12}	31 000
Urease	3.5×10^{-11}	470
Catalase	4.1×10^{-11}	250
Haemoglobin	6.3×10^{-11}	67
Insulin	8.2×10^{-11}	41

$$\Lambda = \sqrt{2\left(5 \times 10^{-12}\,\frac{m^2}{s}\right) 3600\,s} = 1.9 \times 10^{-4}\,m = 0.19\,mm \qquad \textbf{(38)}$$

This performance of the tobacco mosaic virus isn't impressive. It takes 1 hour to travel about 0.2 mm, 4 days to travel about 2 mm and more than 13 months to travel a distance of 2 cm. Clearly, viruses choose other modes of transportation, such as drag effects in air or water, which allow them to spread more quickly.

Solution to part (b): The two formulas for the diffusion lengths read:

$$\Lambda_{O_2} = \sqrt{2 \cdot D_{O_2} \cdot t}$$
$$\Lambda_{CO_2} = \sqrt{2 \cdot D_{CO_2} \cdot t} \qquad \textbf{(39)}$$

and their ratio is given by:

$$\frac{\Lambda_{CO_2}}{\Lambda_{O_2}} = \sqrt{\frac{D_{CO_2}}{D_{O_2}}} = \sqrt{8.3} = 2.9 \qquad \textbf{(40)}$$

in which we have used the result of Problem 10.16 in the form:

$$D_{CO_2} \cong 8.3 \cdot D_{O_2} \qquad \textbf{(41)}$$

This means that carbon dioxide diffuses in tissue roughly three times farther than oxygen under the same conditions.

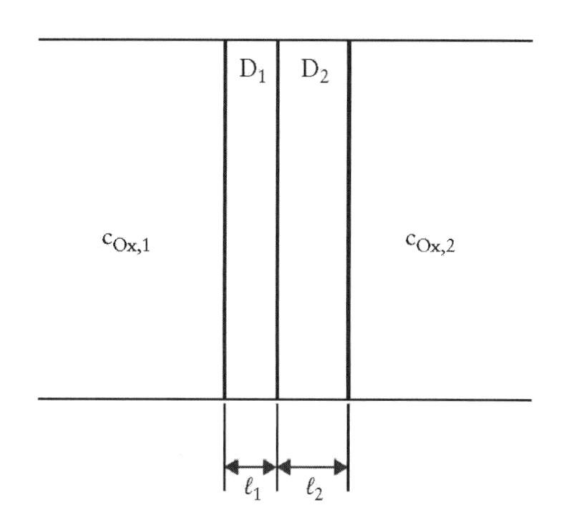

FIGURE 10.37

Problem 10.21

Fig. 10.37 shows a block that consists of two materials with different thicknesses l_1 and l_2 and different diffusion coefficients for oxygen D_1 and D_2. The oxygen concentrations at the outer surfaces of the block are $c_{ox,1}$ and $c_{ox,2}$, as shown in the figure. Each face of the block has a cross–sectional area A. Determine a formula for the steady state mass transport through the block in analogy to Problem 10.7.

Solution: See the solution to Problem 10.7, then write the corresponding equations for the diffusion case.

CHAPTER ELEVEN

Liquid water and aqueous solutions: Static fluids

MULTIPLE CHOICE AND CONCEPTUAL QUESTIONS

Question 11.1
Which law is used to quantify the pressure at the bottom of a lake? (A) Arrhenius' law, (B) Pascal's law, (C) Newton's first law, (D) Laplace's law, (E) none of these.

Answer: (B)

Question 11.3
Which of the following pressures can be negative? (A) the transmural pressure in the lungs, (B) the blood pressure in supine position, (C) the alveolar pressure, (D) the air pressure, (E) the water pressure below the surface of a lake.

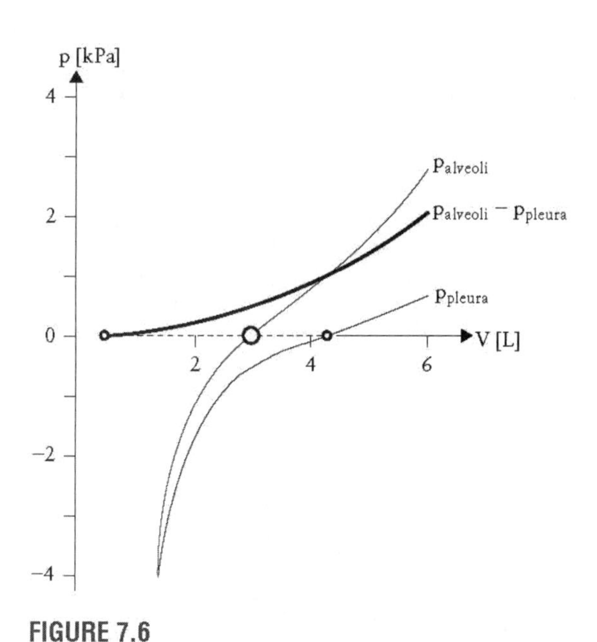

FIGURE 7.6

Answer: (C). Only gauge pressures can be negative. This rules out answers (D) and (E). Figs 7.6 and 7.8 show that the transmural pressure is always positive or zero under normal conditions of the lungs. The alveolar pressure is negative below lung volumes of 3 L for the respiratory curve at rest (Fig. 7.6), and during dynamic inhalation in Fig. 7.8. The blood pressure in supine position is shown in Fig. 11.4 and is discussed in Example 11.2. Negative values do not occur in this position, though they can occur while standing.

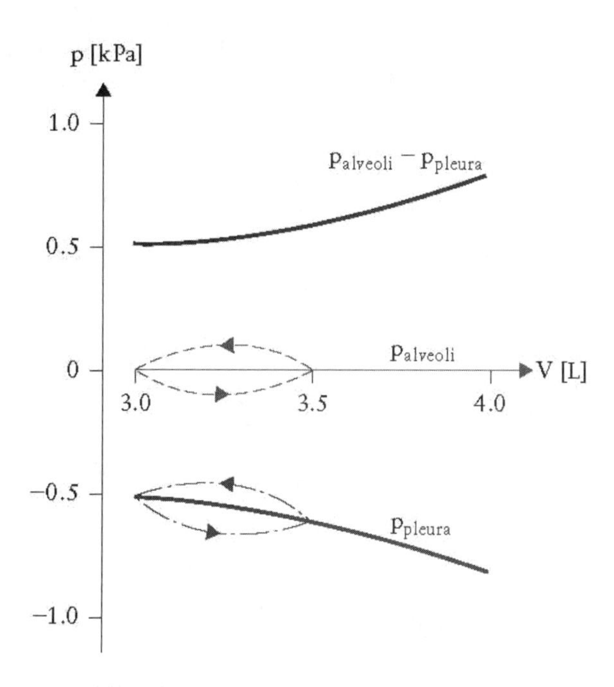

FIGURE 7.8

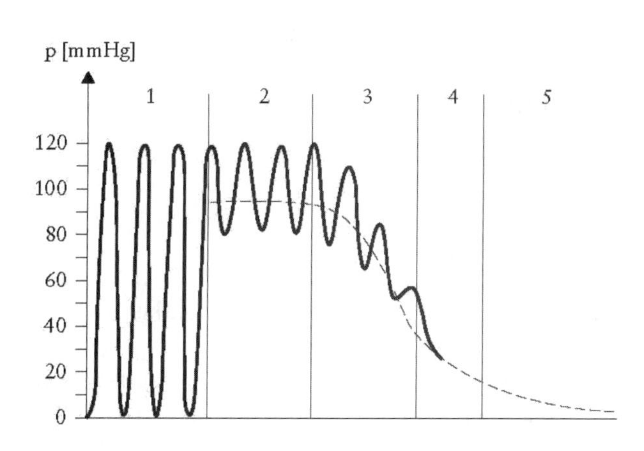

FIGURE 11.4

Question 11.5
Pascal's law is applied to an unknown liquid in an open container. We observe that the pressure is 1.05 atm at 10 cm below the surface. At what depth in the liquid is the pressure 1.2 atm? (A) 0.2 m, (B) 0.4 m, (C) 0.8 m, (D) 1.2 m, (E) 2.5 m.

Answer: (B). In this case, Δp is 0.05 atm per 10 cm. This means that at depth of 40 cm, Δp is 0.2 atm.

Question 11.7
Exchanging one fluid for another in a given beaker, the density of the beaker's content increases if (A) the fluid mass decreases per unit volume, (B) the fluid volume decreases per unit mass, (C) both fluid mass and volume double, (D) both fluid mass and volume are halved.

Answer: (B)

Question 11.9
We consider a motionless shark slowly sinking to the bottom of a lagoon. Which of the choices in Fig. 11.23 represents the proper free–body diagram for the shark of weight **W**? *Hint*: the arrows are not drawn proportional to magnitude of the respective force.

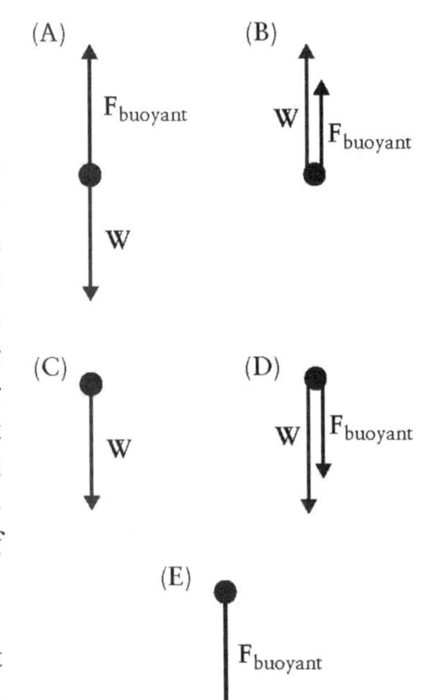

FIGURE 11.23

Answer: (A). Weight must be acting down, and a force must be acting in opposition to that force, otherwise the shark would be falling with an acceleration equal to g.

Question 11.11
The Siamese fighting fish (see Fig. 11.24) have an unusual way of caring for their young. Males prepare a home for their future offspring by taking gulps of air and blowing saliva–coated bubbles that collect at the water's surface as a glistening froth. When a female arrives ready to mate, she swims under the bubble–nest where the male embraces her and fertilizes her eggs. Then, picking up the eggs in his mouth, he spits them, one by one, into the bubbles. The role of father continues for the male Siamese fighting fish as he conscientiously watches over the developing eggs. Any that slip from a bubble and start to sink are carefully retrieved and spat back into a bubble. Based on these observations, what statement can we make about the Siamese fighting fish's fertilized eggs? (A) They are heavy. (B) They have a small volume. (C) Their density is larger than that of water. (D) Their density is equal to that of water. (E) Their density is smaller than that of water.

Answer: (C)

Figure 11.24

Question 11.13
Consider the Hindenburg zeppelin, which exploded in 1937 at Lakehurst, New Jersey. At the time of the accident, its tanks were filled with hydrogen gas. Had they been filled with the following gas, the airship could have carried more passengers: (A) helium, (B) methane, (C) carbon monoxide, (D) nitrogen, (E) none of these.

Answer: (E)

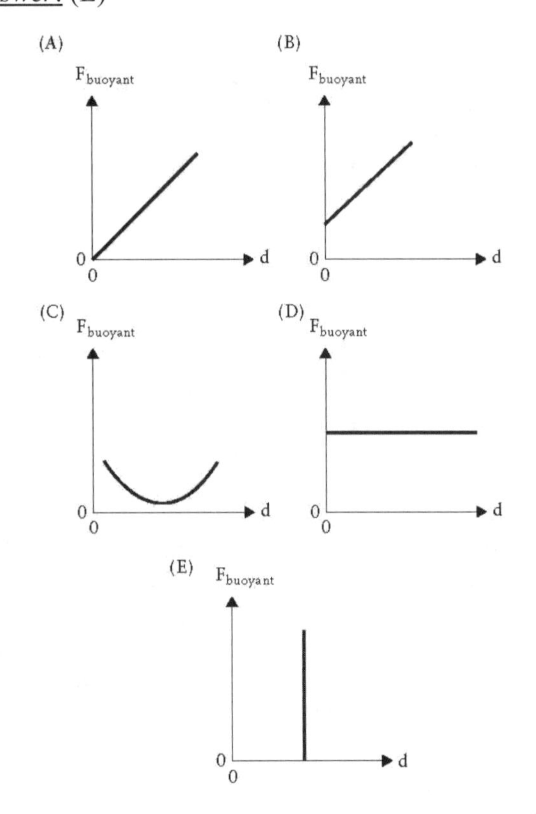

FIGURE 11.26

Question 11.15
We study a solid steel sphere completely immersed in water (which we treat as an ideal stationary fluid). We use the variable d to represent the depth below the water surface. Which sketch in Fig. 11.26 shows the dependence of the buoyant force acting on the solid sphere as a function of depth?

Answer: (D)

Question 11.17
We study Fig. 11.27. Part (a) shows a sphere with a radius $r = 10$ cm and density of $\rho_a = 1.8$ g/cm³ suspended in water. Part (b) shows a wooden sphere of diameter $d = 10$ cm (density $\rho_b = 0.95$ g/cm³) anchored under water with a string. Which of the four free–body diagrams shown in Fig. 11.28 correspond to the two cases in Fig. 11.27? *Note*: **T** is tension, **W** is the weight, and $\mathbf{F}_{\text{buoyant}}$ is buoyant force. (A) The free–body diagram in sketch Fig. 11.28(A) belongs to the case in Fig. 11.27(a) and the free–body diagram in sketch Fig. 11.28(B) belongs to the case in Fig. 11.27(b). (B) The free–body diagram in sketch Fig. 11.28(B) belongs to the case in Fig. 11.27(a) and the free–body diagram in sketch Fig. 11.28(C) belongs to the case in Fig. 11.27(b). (C) The free–body diagram in sketch Fig. 11.28(C) belongs to the case in Fig. 11.27(a) and the free–body diagram in sketch Fig. 11.28(D) belongs to the case in Fig. 11.27(b). (D) The free–body diagram in sketch Fig. 11.28(C) belongs to the case in Fig. 11.27(a) and the free–body diagram in sketch Fig. 11.28(A) belongs to the case in Fig. 11.27(b). (E) The free–body diagram in sketch Fig. 11.28(B) belongs to the case in Fig. 11.27(b) and the free–body diagram in sketch Fig. 11.28(C) belongs to the case in Fig. 11.27(a).

Answer: (E)

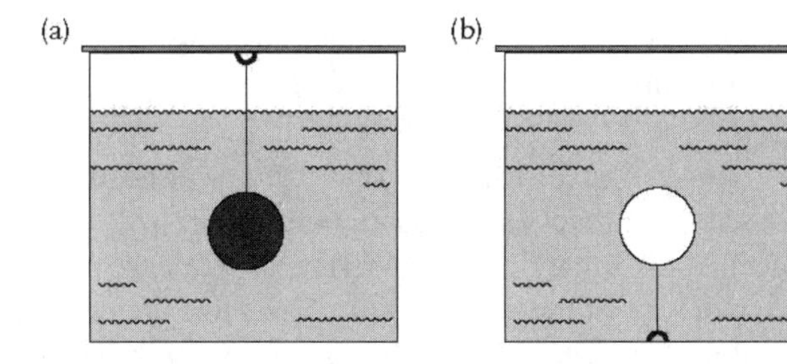

FIGURE 11.27

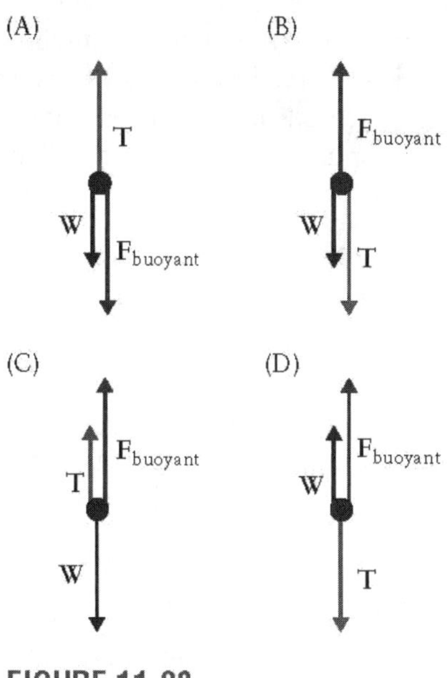

FIGURE 11.28

Question 11.19
Laplace's law describes the pressure in an alveolus in the lungs in the form $p_{\text{inside}} - p_{\text{outside}} = 2 \cdot \sigma/r$, in which σ is the surface tension and r is the radius of curvature of the alveolus. In healthy alveoli a surfactant is used to reduce the surface tension by coating parts of the surface. Which of the following statements is false? (A) The surfactant particularly must coat areas with a large radius of curvature. (B) The surfactant particularly must coat areas with a small radius of curvature. (C) A surfactant does not change the pressure in the bubble.

Answer: (B). The surfactant does not change the pressure in the bubble because adjacent alveoli have connected air space via the bronchial tree. Example 11.5 shows that the surfactant is needed particularly for reducing the surface tension for small alveoli. Small alveoli produce bubbles with a smaller radius of curvature.

Question 11.21
Which law is used to quantify the pressure in a soap bubble? (A) Jurin's law, (B) Pascal's law, (C) Newton's third law, (D) Laplace's law, (E) none of these.

Answer: (D)

Question 11.23
A water–strider is an insect that can walk on water. Looking at one of the six legs of the water–strider resting on the water surface, the following is the case for the

surface underneath the insect's foot: (A) it is perfectly flat, (B) it is bent upward due to the attractive force of the foot, (C) it is curved downward because the foot simulates an increased pressure from above the water surface, (D) it splits and the lower end of the foot dangles below the surface.

Answer: (C)

Question 11.25

A few lizards have developed specialized feet that allow them to climb smooth surfaces, such as windows, without the aid of claws. Examples include several gecko species, which can even walk across the ceiling in tropical homes.

(a) How can they hold on to such a smooth surface?

(b) How can they walk at reasonable pace up the window glass surface?

Answer to part (a): The feet of geckos have enlarged overlapping plates on their undersurfaces, called *lamellae*. The lamellae in turn are covered by so–called *setae*, which are microscopic projections of the skin. These spatula–shaped prongs typically measure 10 μm to 0.1 mm in length. The gecko brings the expanded tips of the setae in close contact with the underlying surface causing weak adhesive forces. These forces are sufficient to hold the gecko at a flat vertical surface because each toe has more than 1 million setae.

The lamellae are pushed against the flat surface with an elaborate mechanism within the gecko's toes. An extensive capillary bed connects to a small blood reservoir beneath the bones in each toe. The gecko can shut off this system from its circulatory system with a set of valves. When the bone then pushes onto the filled blood reservoir, the capillaries are pressurized. The blood vessels expand and push onto the adjacent lamellae. This forces the lamellae tightly against the surface, following any of its irregularities closely.

Since the surface energy of the substrate is a factor in the adhesive force, geckos find smooth glass surfaces with their large surface energy particularly easy to hold onto.

Answer to part (b): The adhesion achieved in part (a) poses a problem when the gecko wants to walk. To lift the foot off the surface, the gecko must depressurize the capillary bed. To detach from the surface, the toe is rolled off from tip to base to not have to peel all setae off the surface at once. These attachment and detachment processes have to occur for every step taken.

ANALYTICAL PROBLEMS

Problems 11.1

The sphere is the shape with the smallest surface for a given volume. To prove this statement properly requires variational analysis. Here we only want to confirm this result for a selection of highly symmetric shapes by calculating the ratio of surface and volume. Find these ratios for (a) sphere, (b) cylinder, (c) cube, (d) pyramid, (e) tetrahedron, and (f) cone. Does the statement hold for these six shapes? Use the Math Review *Symmetric Objects* on page 82 in Chapter 3 for some required data.

Solution: We use a sphere of radius r, a cylinder of height h and radius r with $h = r$, a cube of side length h, a pyramid of base length h with four equilateral triangles, a tetrahedron of side length h, a cone with base radius r and height h with $r = h$, and an octahedron which is a double–pyramid. For these bodies, Table 11.2 shows a sketch, the formula for the volume, the value for r or h when the volume is $V = 1\ \text{m}^3$, the formula for the surface and the total surface area in unit m^2. Note that the sphere indeed has the smallest surface of all of these bodies of equal volume.

Problem 11.3

A scuba diver takes a deep breath from an air–filled tank at depth d, then abandons the tank. During the subsequent ascent to the surface the diver fails to exhale. When reaching the surface, the pressure difference between the external pressure and the pressure in the lungs is 76 torr. At what depth did the diver abandon the tank? *For those interested*: What potentially lethal danger does the diver face?

Solution: This is an example of a possible diving accident when scuba diving. *Scuba* stands for *self–contained under water breathing apparatus*, a technique developed by Jacques–Yves Cousteau in 1943.

When filling the lungs at depth d, the external pressure on the body and, correspondingly, the gas pressure in the lungs is given by Pascal's law, i.e., $p = p_{\text{atm}} + \rho \cdot g \cdot d$.

As the diver ascends, the external pressure decreases until it reaches atmospheric pressure p_{atm} at the surface. At the same time the blood pressure decreases to its normal (out of the water) value. But the gas pressure in the lungs remains at the value it was at depth d since the diver doesn't exhale. Therefore, the pressure difference between the lungs and the outside pressure on the chest at the surface is:

Body	Sketch	Volume Formula	r, h for V = 1	Surface Formula	Surface for V = 1
sphere		$\frac{4}{3}\pi r^3$	0.6204	$4\pi r^2$	4.8
cylinder (r = h)		$r^2\pi h$	0.6828	$2r\pi(r + h)$	5.9
cube		h^3	1.0	$6\,h^2$	6.0
pyramid		$\frac{h^3}{3\sqrt{2}}$	1.6189	$h^2\,(1 + \sqrt{3})$	5.4
tetrahedron		$\frac{h^3}{12}\sqrt{2}$	2.0396	$h^2\sqrt{3}$	7.2
cone (r = h)		$\frac{r^2}{3}\pi h$	0.9847	$r\pi(r+\sqrt{r^2+h^2})$	7.4
octahedron		$\frac{h^3}{3}\sqrt{2}$	1.2849	$2h^2\sqrt{3}$	5.7

Table 11.2

$$\Delta p = p - p_{atm} = \rho \cdot g \cdot d \quad \textbf{(1)}$$

and the depth d is given by:

$$d = \frac{\Delta p}{\rho \cdot g} = \frac{(76\ torr)\left(133.32\ \frac{Pa}{torr}\right)}{\left(1.0 \times 10^3\ \frac{kg}{m^3}\right)\left(9.8\ \frac{m}{s^2}\right)} \quad \textbf{(2)}$$

$$= 1.03\ m$$

in which we used the pressure conversion of 1 torr = 133.32 Pa. Thus, surfacing from about a 1 m depth without exhaling in this case leads to 76 torr pressure difference, which is about 10% of the atmospheric pressure. It is still enough to rupture the lungs and force air from the higher pressure lungs into the lower pressure blood. This air may be carried to the heart and kill the diver. An ugly way to die!

Problem 11.5

What minimum gauge pressure is needed to suck water up a straw to a height of 10 cm? Recall that the gauge pressure is defined as the pressure relative to atmospheric pressure, $p_{gauge} = p - p_{atm}$.

Solution: This problem is solved using Pascal's law. Using the definition of the gauge pressure from the problem text, Pascal's law is rewritten in the form:

$$p_{gauge} = p - p_{atm} = -\rho \cdot g \cdot h \qquad (3)$$

in which the negative sign on the right hand side indicates that the pressure in the lungs is lower than the atmospheric pressure during the sucking. We substitute the numerical values in Eq. [3]:

$$p_{gauge} = -\frac{\left(1.0 \times 10^3 \frac{kg}{m^3}\right)\left(9.8 \frac{m}{s^2}\right)(0.1\ m)}{1.01 \times 10^5 \frac{Pa}{atm}} \qquad (4)$$

$$= -9.7 \times 10^{-3}\ atm$$

The denominator in Eq. [4] allows us to convert to the non–standard unit atm. Of course, the best way to answer the question would be to provide an answer in the standard unit Pa; however, non–standard pressure units are still quite often used.

Problem 11.7

Water is pumped to the top of the 365-m-tall Empire State Building in New York City. What gauge pressure is needed in the water line at the base of the building to achieve this?

Solution: $p = \rho gh = (1 \times 10^3\ kg/m^3)(9.8\ m/s^2)(365\ m) = 3577\ kPa$.

Problem 11.9

The density of ice is $\rho_{ice} = 920\ kg/m^3$ and the average density of seawater is $\rho_w = 1.025\ g/cm^3$. What fraction of the total volume of an iceberg is exposed?

Solution: For this problem we use the Archimedes principle. The magnitude of the weight of an iceberg of total volume V_{total} is:

$$W_{total} = m_{ice} \cdot g = \rho_{ice} \cdot V_{total} \cdot g \qquad (5)$$

The magnitude of the weight of the displaced seawater is equal to the magnitude of the buoyant force acting on the iceberg, $F_{buoyant}$:

$$F_{buoyant} = m_{H_2O} \cdot g = \rho_{H_2O} \cdot V_{displaced\ water} \cdot g \qquad (6)$$

in which $V_{displaced\ water}$ is equal to the volume of the iceberg below the surface of the sea.

Figure 11.32 ⇒

$\vec{F}_B$

$\vec{W}_{tot}$

The iceberg is in mechanical equilibrium when it floats, i.e., the weight of the iceberg and the buoyant force must be equal, as illustrated in the free–body diagram in Fig. 11.32. Using Newton's first law we write for the balance of forces in the vertical direction:

$$\rho_{H_2O} \cdot V_{displaced\ water} \cdot g - \rho_{ice} \cdot V_{total} \cdot g = 0 \qquad (7)$$

which yields:

$$\frac{\rho_{ice}}{\rho_{H_2O}} = \frac{V_{displaced\ water}}{V_{total}} \qquad (8)$$

where in this case ρ_{H20} refers to the density of sea water. This provides us with a formula for the ratio of the volume of the displaced water to the total volume of the iceberg. We now express the quantity sought in the problem. The question is about the fraction of the volume of the iceberg which reaches above the sea level:

$$\frac{V_{total} - V_{displaced\ water}}{V_{total}} = 1 - \frac{V_{displaced\ water}}{V_{total}} \qquad (9)$$

We use Eq. [8] to replace the second term on the right hand side:

$$1 - \frac{V_{displaced\ water}}{V_{total}} = 1 - \frac{\rho_{ice}}{\rho_{H_2O}}$$

$$= 1 - \frac{0.92 \frac{kg}{L}}{1.025 \frac{kg}{L}} = 0.102 \qquad (10)$$

For icebergs off the coast of Newfoundland and Labrador, just over 10% of the iceberg volume extends beyond the surface of the water. Hence the expression "the tip of the iceberg", which is used to indicate that there is much more to something that is apparent.

Problem 11.11

(a) Fig. 11.27(b) shows a wooden sphere with a diameter of $d = 10$ cm (density $\rho = 0.9$ g/cm³) held under water by a string. What is the tension in the string?

(b) Fig. 11.27(a) shows a sphere of radius $r = 10$ cm and density of $\rho = 2.0$ g/cm³ suspended in water. What is the tension in the sting? *Note*: draw the free–body diagram in each case.

Solution to part (a): An object which floats at a given depth below the surface of a fluid must be in mechanical equilibrium. For the three cases we considered in the context of Fig. 11.7, this was discussed as case (II). In Fig. 11.7, such a mechanical equilibrium is based on two forces acting on the block B: the weight and the buoyant force. The situation in Fig. 11.27 is different because in both cases an additional tension force acts on the object. Thus, without the string, neither case 11.27(a) nor case 11.27(b) would be in a mechanical equilibrium. It is the additional tension which establishes this equilibrium. Consequently, the free–body diagram in each case must be based on three forces.

The free–body diagram for the wooden sphere is shown in Fig. 11.28(B). The buoyant force is drawn with a larger magnitude than the weight because we know from experience that the wooden sphere, if released, would buoy to the surface. To establish mechanical equilibrium a second downward directed force is needed. This force is provided by the tension in the string.

For mechanical equilibrium, the free–body diagram must show the length of the tension force to be equal to the difference in length between the weight and the buoyant force (unlike in Fig. 11.28(B)). Using Newton's first law, we write:

$$F_{buoyant} - T - W = 0 \qquad \textbf{(11)}$$

Of the three forces in Eq. [11] we can quantify two:

- the weight: Use the volume of the wooden sphere and its density. Using the radius $r = d/2 = 5$ cm, we get:

$$\begin{aligned} W &= m_{wood} \cdot g = V \cdot \rho_{wood} \cdot g \\ &= \frac{4}{3}\pi \cdot r^3 \cdot g \cdot \rho_{wood} \end{aligned} \qquad \textbf{(12)}$$

- the buoyant force:

$$F_{buoyant} = V_{H_2O} \cdot \rho_{H_2O} \cdot g = \frac{4}{3}\pi \cdot r^3 \cdot g \cdot \rho_{H_2O} \qquad \textbf{(13)}$$

Now we substitute these two forces in Eq. [11] and solve for the magnitude of the tension **T**:

$$T = F_{buoyant} - W = \frac{4}{3}\pi \cdot r^3 \cdot g\left(\rho_{H_2O} - \rho_{wood}\right) \qquad \textbf{(14)}$$

Substituting the given values the right hand side reads:

$$\frac{4}{3}\pi\,(0.05\,m)^3\left(9.8\,\frac{m}{s^2}\right)\left(1.0\,\frac{g}{cm^3} - 0.9\,\frac{g}{cm^3}\right) \qquad \textbf{(15)}$$

This yields $T = 0.51$ N (remember to change the density into units of kg/m³).

Solution to part (b): The free–body diagram for the heavy sphere from Fig. 11.27(a) is shown in Fig. 11.28(C). The string provides for a mechanical equilibrium. The tension associated with the sting is calculated from applying Newton's first law:

$$F_{buoyant} + T - W = 0 \qquad \textbf{(16)}$$

Note that this equation differs from Eq. [11] for case (a) only in the sign of the tension as the string is now directed upward. Again, the buoyant force and the weight of the heavy sphere can be calculated; the formulas are identical to those shown in Eqs. [12] and [13] if we replace the label *wood* by *heavy sphere*. Note that the radius of the sphere is given in part (b) instead of its diameter. Substituting Eqs. [12] and [13] in Eq. [16] and solving for the magnitude of the tension **T**, we obtain:

$$T = W - F_{buoyant} = \frac{4}{3}\pi \cdot r^3 \cdot g\left(\rho_{heavy} - \rho_{H_2O}\right) \qquad \textbf{(17)}$$

This yields $T = 41.05$ N.

Problem 11.13

Fig. 11.31(a) shows how surface tension supports insects such as water–striders on the water surface. Assume that an insect's foot is spherical as shown in Fig. 11.31(b), and that the insect stands with all of its six feet on the water. Each foot presses the water surface down while the surface tension of the water produces upward forces to restore the normal flat shape of the water surface. A characteristic profile of the water surface results, as shown in the figure. The mass of the insect is 15 mg and

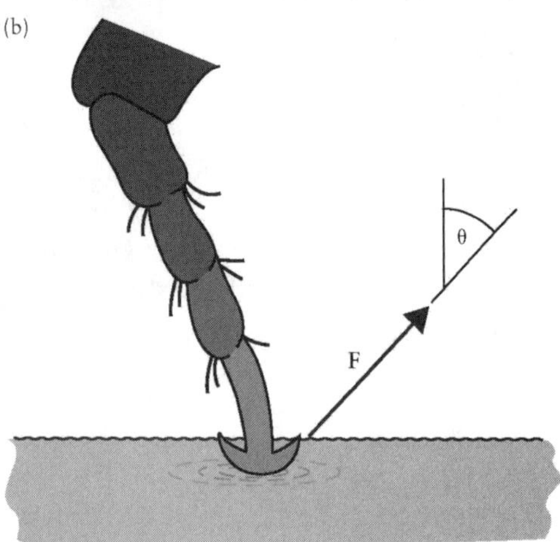

FIGURE 11.31

the diameter of the insect's foot is 250 μm. Find the angle θ as indicated in Fig. 11.31(b). *Hint*: The definition of the surface tension provides for a tangential force along the depressed surface of the water, shown as force **F** in the figure. The surface tension of water at 20°C is σ = 0.073 N/m.

Solution: The definition of surface tension used for this problem is given in Eq. [10.30]:

$$\sigma = \frac{F}{l_x} \quad \textbf{(18)}$$

Note that the force in that formula is directed *tangential* to the depressed water surface, as properly indicated in Fig. 11.31. The length l_x in Eq. [18] represents the length of the contact line along which the force of magnitude F acts. In the case of the insect's foot, this line is the circumference of the foot where it touches the surface of the water. This circumference is determined from the given diameter of the insect's foot by using Fig. 11.33. The figure is a detailed view of the geometric relations of the force and spatial parameters relevant to this problem. We note that the circumference of the foot at the water line is given by $2 \cdot \pi \cdot r \cdot \cos\theta$ in which r is the radius of the foot. We label this circumference l_x.

Using the value for l_x, we find from Eq. [18] the magnitude of the force **F** exerted by the water surface, as illustrated in Figs. 11.31 and 11.33:

$$F = \sigma \cdot l_x = \sigma \cdot 2 \cdot \pi \cdot r \cdot \cos\theta \quad \textbf{(19)}$$

Next we use Newton's first law to describe the mechanical equilibrium of the insect. Of interest is only the vertical direction as the symmetry of the foot provides automatically for a compensation of the *x*–directional components of the forces acting on the foot. The equilibrium must exist because the insect neither accelerates down (and drowns) nor accelerates up (while not intentionally flying away). Since an insect has six legs and usually rests with all of them on the water surface, Newton's law contains seven contributions: one force in the positive *y*–direction due to the force **F** acting on each leg, and one force down due to the weight of the insect:

$$-W_{insect} + 6\,(\sigma \cdot 2 \cdot \pi \cdot r \cdot \cos\theta)\cos\theta = 0 \quad \textbf{(20)}$$

in which the extra cosθ term is introduced by using the *y*–component of the force **F** in Fig. 11.33. We solve Eq. [20] for $\cos^2\theta$:

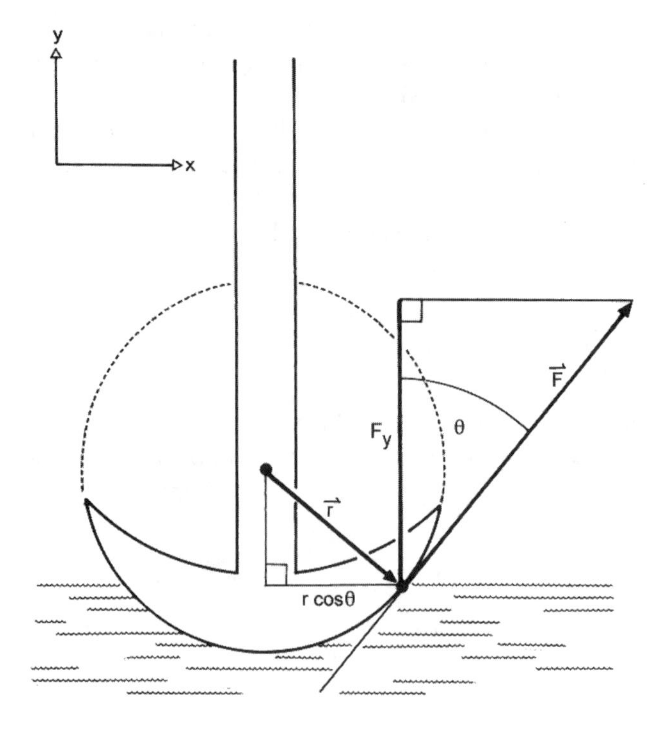

Figure 11.33

$$\cos^2\theta = \frac{m \cdot g}{12 \cdot \pi \cdot r \cdot \sigma}$$

$$= \frac{1.5 \times 10^{-5}\, kg \left(9.8\, \frac{m}{s^2}\right)}{12 \cdot \pi \cdot (1.25 \times 10^{-4}\, m) \left(0.073\, \frac{N}{m}\right)} \qquad \textbf{(21)}$$

$$= 0.427$$

in which m is the mass of the insect. Thus, we obtain $\cos\theta = 0.653$ which corresponds to an angle $\theta = 49^0$. This illustrates that contrary to our intuition the depression of the water is rather steep, with a 41^0 angle between the flat water surface and the edge of the foot.

CHAPTER TWELVE

Cardiovascular system: Fluid flow

MULTIPLE CHOICE AND CONCEPTUAL QUESTIONS

Question 12.1
The volume flow rate and the mass flow rate in laminar flow are (A) the same, (B) proportional to each other, (C) inversely proportional to each other, (D) unrelated, (E) related in a non–linear fashion.

Answer: (B)

Question 12.3
Fig. 12.31 shows a cylindrical tube of changing diameter with an ideal dynamic fluid (grey) flowing toward the right with initial speed v. The vertical columns are connected to the main tube. Which of the five choices shows the proper elevations of the fluid in each of the three vertical columns?

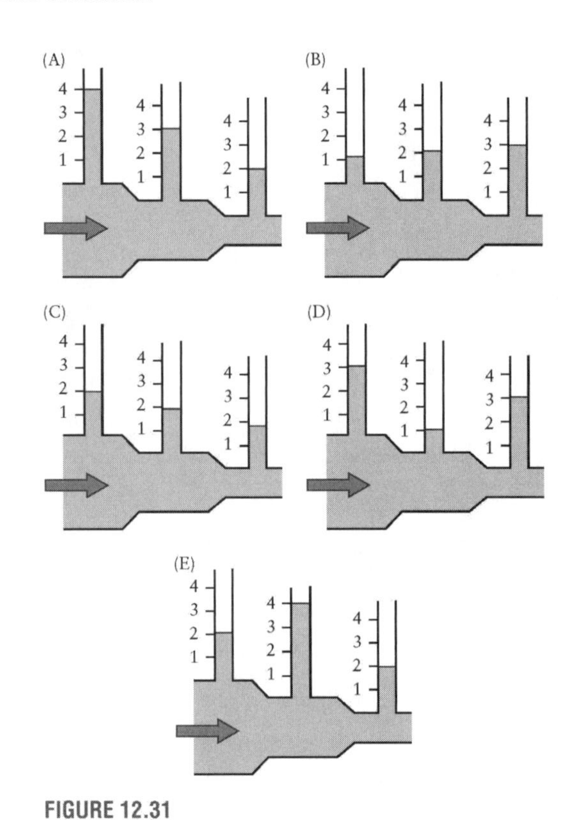

FIGURE 12.31

Answer: (A). Smaller cross–section means faster fluid flow (equation of continuity), which, in turn, means lower fluid pressure (Bernoulli's law). Low pressures occur closer to the surface in static fluids (Pascal's law).

Question 12.5
The equation of continuity is an expression of (A) the conservation of mass, (B) the conservation of total energy, (C) the conservation of kinetic energy, (D) the conservation of velocity.

Answer: (A)

Question 12.7
Which law connects the speed of an ideal dynamic fluid to its pressure? (A) Pascal's law, (B) Newton's second law, (C) Bernoulli's law, (D) Laplace's law, (E) none of these.

Answer: (C)

Question 12.9
The conservation of mass leads to the following law we use to describe laminar flow in a fluid: (A) Bernoulli's law, (B) Pascal's law, (C) equation of continuity, (D) Poiseuille's law, (E) Ohm's law.

Answer: (C)

Question 12.11
Do the following experiment: hold two sheets of paper parallel to each other at a distance of about 1 cm to 3 cm. Blow gently between the two sheets from their edges. The two sheets will be pulled together. Which law explains this observation? (A) Poiseuille's law, (B) Ohm's law, (C) equation of continuity, (D) Bernoulli's law, (E) Pascal's law.

Answer: (D)

Question 12.13
An artery is partially clogged by a deposit on its inner wall, as shown in Fig. 12.33. Which of the following statements best describes the processes that occur when

blood rushes through this constriction? Treat blood as a Newtonian fluid. (A) Blood will rush faster through the constriction due to the equation of continuity, causing additional wear and tear on the nearby blood vessel walls. (B) Bernoulli's law and the equation of continuity predict a variation of the blood pressure in the constricted zone, but the blood vessel walls prevent any adverse effect due to this pressure variation. (C) The blood pressure in the constriction zone is lower than in the adjacent blood vessel, causing the blood vessel to temporarily collapse at the constriction (vascular flutter). (D) The blood pressure in the constriction zone is higher than in the adjacent blood vessel, causing a ballooning effect of the blood vessel at the constriction (aneurysm).

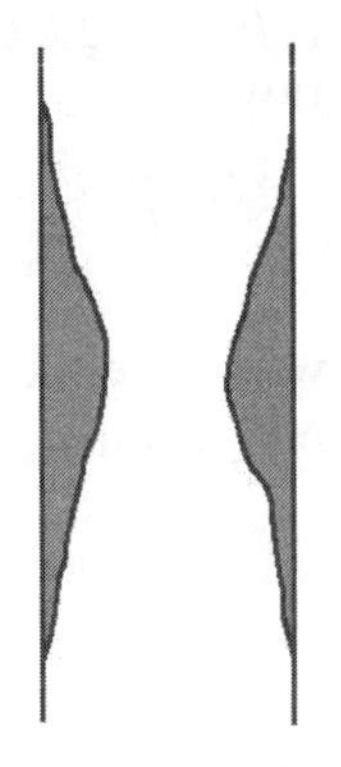

FIGURE 12.33

Answer: (C)

Question 12.15

North American prairie dogs live in underground burrows with several exits. They usually build a mound over one exit which causes a draft past that hole. How does this arrangement allow for ventilation of the burrow with the air stagnant above all other exits?

Answer: Air passing over the mound speeds up while passing over the hole. This lowers the air pressure above the hole, causing suction on the air in the burrow.

Question 12.17

Tornados and hurricanes can lift roofs off houses. A standard recommendation to home owners in affected areas is to keep windows open when a storm approaches. What happens to the roof, and why would open windows help?

Answer: The roof can be compared to a wing. Above, wind passes with high speed (and corresponding low pressure) while the air below the roof is stationary and at regular air pressure. This pressure difference causes an upward force on the roof. Open windows help to lower the pressure inside the house, i.e., equilibrate the pressure inside and outside.

Question 12.19

We study flow of a Newtonian fluid through two different tubes (index 1 and 2). The pressure differences between the two end of the tubes, Δp, are the same for both tubes, $\Delta p_1 = \Delta p_2$. The tubes differ in radius and length: Length of tube 1 is $l_1 = 2$ m, length of tube 2 is $l_2 = 1$ m, radius of tube 1 is $r_1 = 2$ cm, and radius of tube 2 is $r_2 = 1$ cm. Which of the following is the correct ratio of volume flow rates through the two tubes? (A) $\Delta V/\Delta t_1 : \Delta V/\Delta t_2 = 2:1$. (B) $\Delta V/\Delta t_1 : \Delta V/\Delta t_2 = 4:1$. (C) $\Delta V/\Delta t_1 : \Delta V/\Delta t_2 = 8:1$. (D) $\Delta V/\Delta t_1 : \Delta V/\Delta t_2 = 16:1$. (E) none of the above.

Answer: (C)

Question 12.21

Which of the following statements is wrong? (A) Poiseuille's law applies as derived only to laminar flow. (B) Poiseuille's law applies as derived only to ideal dynamic fluids. (C) Poiseuille's law applies as derived only to incompressible fluids. (D) Poiseuille's law and the equation of continuity can be used together for the same system. (E) Poiseuille's law applies as derived only to Newtonian fluids.

Answer: (B)

ANALYTICAL PROBLEMS

Problem 12.1

What is the net upward force on an airplane wing of area $A = 20\ \text{m}^2$ if air streams at 300 m/s across its top and at 280 m/s past the bottom. Note that this airplane moves at subsonic speed with respect to the speed of sound (called Mach 1), about 330 m/s.

Solution: We use Bernoulli's law:

$$p + \frac{1}{2}\rho \cdot v^2 = const \quad \textbf{(1)}$$

at the top and bottom of the airplane wing in the form:

$$p_{top} + \frac{1}{2}\rho_{air} \cdot v_{top}^2 = p_{bottom} + \frac{1}{2}\rho_{air} \cdot v_{bottom}^2 \quad \textbf{(2)}$$

where $\rho_{air} = 1.2\ \text{kg/m}^3$ is the density of air. We re–arrange Eq. [2]:

$$\Delta p = p_{bottom} - p_{top} = \frac{1}{2}\rho_{air}\left(v_{top}^2 - v_{bottom}^2\right) \quad \textbf{(3)}$$

This yields:

$$\Delta p = \frac{1}{2}\left(1.2\,\frac{\text{kg}}{\text{m}^3}\right)\left\{\left(300\,\frac{\text{m}}{\text{s}}\right)^2 - \left(280\,\frac{\text{m}}{\text{s}}\right)^2\right\} \tag{4}$$
$$= 6960\,\frac{\text{N}}{\text{m}^2}$$

This pressure difference causes a net upward force on the wing of:

$$F_{net} = \Delta p \cdot A = \left(6960\,\frac{\text{N}}{\text{m}^2}\right) \cdot (20\ \text{m}^2) \tag{5}$$
$$= 1.4 \times 10^5\ \text{N}$$

Problem 12.3
A large water–containing tank is open to air. It has a small hole 16 m below the water surface through which water leaks at a rate of 2.5 L/min. Determine
(a) the speed of the water that is ejected from the hole, and
(b) the diameter of the hole.

Solution to part (a): We use Bernoulli's law (Eq. [1]). At the location of the small hole, both on the inside and the outside of the tank, we find:

$$p_{in} + \frac{1}{2}\rho \cdot v_{in}^2 = p_{out} + \frac{1}{2}\rho \cdot v_{out}^2 \tag{6}$$

Since the tank is open to the air at the top, we further know from Pascal's law:

$$p_{in} = p_{atm} + \rho \cdot g \cdot h \tag{7}$$

where h is the distance of the hole below the surface of the water. Inside the tank, the fluid speed is approximately zero. Outside the tank we know that:

$$p_{out} = p_{atm} \tag{8}$$

We substitute Eqs. [7] and [8] in Eq. [6]:

$$p_{atm} + \rho \cdot \rho \cdot h = p_{atm} + \frac{1}{2}\rho \cdot v_{out}^2 \tag{9}$$

which simplifies and can be re–arranged as:

$$v_{out} = \sqrt{2\,g\,h} = \sqrt{2\left(9.8\,\frac{\text{m}}{\text{s}^2}\right)(16\ \text{m})} = 18\,\frac{\text{m}}{\text{s}} \tag{10}$$

Solution to part (b): We use the definition of the volume flow rate:

$$\frac{\Delta V}{\Delta t} = A \cdot v \tag{11}$$

with v the speed of the fluid. Re–arranging and converting the units, we obtain:

$$A = \frac{1}{v}\frac{\Delta V}{\Delta t} = 2.3 \times 10^{-6}\ \text{m}^2 \tag{12}$$

Since $A = \pi \cdot r^2 = \pi \cdot (d/2)^2$, we find that $d = 0.0017$ m = 1.7 mm.

Problem 12.5
Water is supplied to a building through a pipe of radius $R = 3.0$ cm. In the building, a faucet tap of radius $r = 1.0$ cm is located 2.0 m above the entering pipe. When the faucet is fully open, it allows us to fill a 25-L bucket in 0.5 minutes.
(a) With what speed does the water leave the faucet?
(b) What is the gauge pressure in the pipe entering the building? Assume that no other faucets are opened during the experiment.

Solution to part (a): We use the definition of the volume flow rate in Eq. [11]. The area of the faucet is calculated from its radius. We solve Eq. [11] for the fluid speed and substitute all values in SI units: $v = 2.7$ m/s.

Solution to part (b): Since there are no other faucets open, the volume flow rate out of the faucet must be equal to the flow rate into the pipe. The cross–sectional area of the inflow pipe is determined from its radius. We can then find the speed of water into the pipe. In the final step, Bernoulli's law is used to calculate $p_{gauge} = 23$ kPa.

Problem 12.7
Fig. 12.36 shows a siphon. Flow in this device must be initiated with suction, but then proceeds on its own.

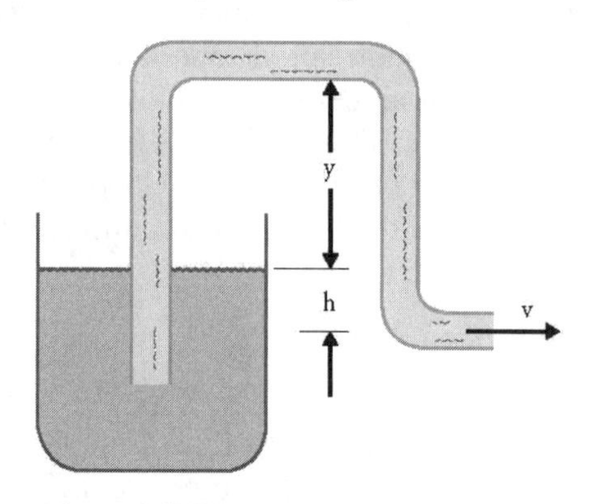

FIGURE 12.36

(a) Show that water emerges from the open end at a speed of $v = (2 \cdot g \cdot h)^{1/2}$.
(b) For what range of y–values will this device work?

Solution to part (a): The open end of the siphon is a distance h below the open surface of the tank. Therefore, the pressure in the fluid as it emerges from the siphon's open end is calculated with Pascal's law. We then use Bernoulli's law comparing the fluid at the surface of the tank and the fluid leaving the open end of the siphon. We note that the pressure at the surface of the tank is atmospheric, and that the fluid at the surface of the tank is essentially at rest. Substituting these conditions into Bernoulli's law leads as an intermediate formula to:

$$p_{atm} = p_{atm} + \rho \cdot g \cdot h + \frac{1}{2} \rho \cdot v_{end}^2 \qquad \textbf{(13)}$$

Solution to part (b): $y_{max} = p_{atm}/(\rho \cdot g)$

Problem 12.9
A beaker has a hole of radius $r = 1.75$ mm near its bottom from which water is ejected as shown in Fig. 12.38. Calculate the height h of the water in the beaker if $h_1 = 1.0$ m and $h_2 = 0.6$ m.

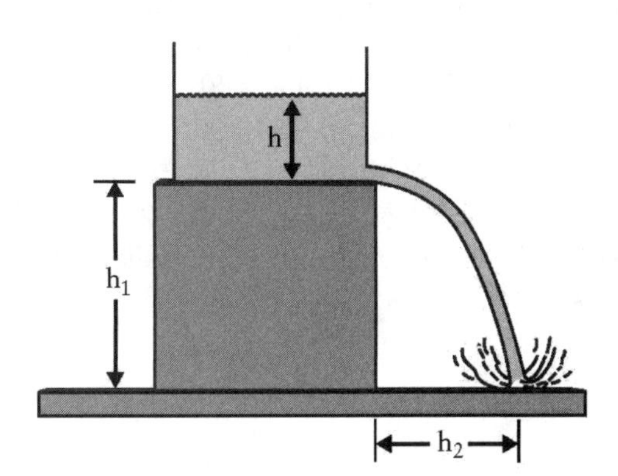

FIGURE 12.38

Solution: Since the beaker is open to the atmosphere at the top, we could use Pascal's law to find the height of water in the beaker if we knew the pressure at the bottom. Bernoulli's law would allow us to find the pressure if we knew the speed of the water as it leaves the hole in the beaker. The speed of the water as it emerges from the hole in the beaker *can* be found because the water follows a projectile trajectory as it travels toward the floor. We use the kinematics relationships for two–dimensional motion in the x–y plane noting that $v_{initial,\,y} = 0$ since the initial velocity is only horizontal. We choose the origin of the system to be the location of the floor, directly below the hole in the beaker. Therefore the initial position of the water is (0.0 m, 1.0 m) and the final position as it hits the floor is (0.6 m, 0.0 m). Substituting the given values in the kinematics equation for the y–direction, we find for the time to the ground $t = 0.45$ s.

This time can be used in the kinematics equation for the x–direction, noting as always that the x–component of acceleration is zero for a projectile. We find $v_{initial,\,x} = 1.33$ m/s.

We use Bernoulli's equation and Pascal's law together comparing the fluid inside and just outside the hole in the beaker, noting the fluid inside the beaker is approximately at rest. This yields $h = 0.09$ m = 9.0 cm.

Problem 12.11
Air moves through the human trachea at 3 m/s during inhalation. Assume that a constriction in the bronchus exists at which the speed doubles. Treating air as an ideal dynamic fluid, calculate the pressure in the constriction.

Solution: Using Bernoulli's law we find that the pressure change in the constriction is $\Delta p = -17$ Pa. This is a very small decrease in pressure in the bronchus.

Problem 12.13
The hypodermic syringe in Fig. 12.39 contains water. The barrel of the syringe has a cross–sectional area $A_1 = 30\ mm^2$. The pressure is 1.0 atm everywhere while no force is exerted on the plunger. When a force $\mathbf{F}_{ext}$ of magnitude 2.0 N is exerted on the plunger, the water squirts from the needle. Determine the water's flow speed through the needle, v_2. Assume that the pressure in the needle remains at a value of $p_2 = 1.0$ atm and that the syringe is held horizontal. The final speed of the water in the barrel is negligible.

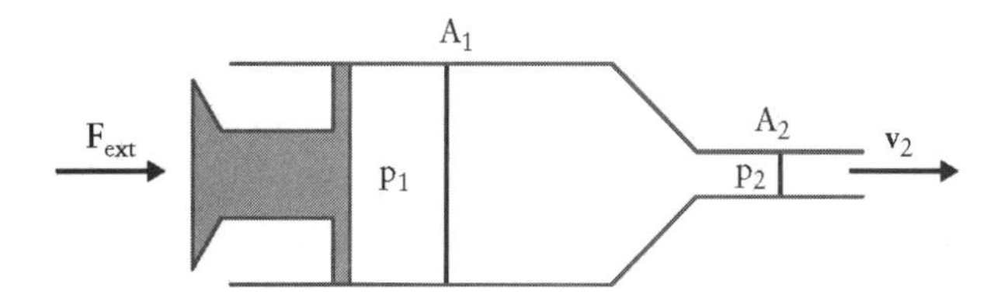

FIGURE 12.39

Solution: We first calculate the pressure of the water in the syringe's barrel, p_1. This value is determined from the mechanical equilibrium of forces that act on the plunger. A mechanical equilibrium must exist since the question asks for a flow speed of the water, implying that there is no acceleration of the plunger involved. In the equilibrium the pressure inside the barrel is equal to the two components acting on the plunger from outside, the atmospheric pressure and the pressure caused by the exerted force:

$$p_1 = \frac{|\boldsymbol{F}_{ext}|}{A_1} + p_{atm} = \frac{2.0\ \text{N}}{3 \times 10^{-5}\ \text{m}^2} + 1.01 \times 10^5\ \text{Pa} = 1.68 \times 10^5\ \text{Pa} \qquad (14)$$

Now we apply Bernoulli's law across the needle of the syringe:

$$p_1 + \frac{\rho_{H_2O}}{2} v_{barrel}^2 = p_2 + \frac{\rho_{H_2O}}{2} v_2^2 \qquad (15)$$

in which we use for the speed of the water in the barrel $v_{\text{barrel}} \cong 0$ m/s as an approximation, and further that $p_2 =$ 1.0 atm for the pressure in the needle. With these simplifications we write:

$$\frac{\rho_{H_2O}}{2} v_2^2 = p_1 - p_{atm} \qquad (16)$$

From this equation the speed of the water in the needle is obtained:

$$v_2 = \sqrt{\frac{2(p_1 - p_{atm})}{\rho_{H_2O}}} \qquad (17)$$

Substituting the given values, we find:

$$v_2 = \sqrt{\frac{2\,(1.68 \times 10^5\ \text{Pa} - 1.01 \times 10^5\ \text{Pa})}{1.0 \times 10^3\ \frac{\text{kg}}{\text{m}^3}}} = 11.6\ \frac{\text{m}}{\text{s}} \qquad (18)$$

Problem 12.15
A horizontal tube of radius $r = 5.0$ mm, and length $l =$ 50 m carries oil ($\eta = 0.12$ N · s/m²). At the end of the tube the flow rate is 85 cm³/s, and the pressure is $p =$ 1.0 atm. What is the gauge pressure the beginning of the tube?

Solution: Poiseuille's law allows us to relate the volume flow rate of a Newtonian fluid through a tube to the pressure gradient in the tube. We start by changing the units of the volume flow rate to SI units. We then rewrite Poiseuille's law to solve for the change in pressure over the tube's length: $p = 2.08 \times 10^6$ Pa = 20.5 atm. If the pressure at the end of the tube is 1 atm, at its beginning we find $p = 2.08 \times 10^6$ Pa + 1.01×10^5 Pa = 2.18×10^6 Pa. But we are asked for the gauge pressure which is the pressure relative to atmospheric, so the gauge pressure is actually $p = 2.08 \times 10^6$ Pa = 20.5 atm.

CHAPTER THIRTEEN

The water molecule: Static electricity

MULTIPLE CHOICE AND CONCEPTUAL QUESTIONS

Question 13.1
Water consists of (A) positively charged atoms, (B) negatively charged atoms, (C) positively charged molecules, (D) negatively charge molecules, (E) none of the above.

Answer: (E)

Question 13.3
Two sodium ions in an aqueous solution (A) repel each other, (B) do not interact at any separation distance, (C) attract each other, (D) seek each other to form metallic sodium precipitates, (E) do none of the above.

Answer: (A). A single positive charge remains unbalanced even after shielding due to a hydration shell.

Question 13.5
The water molecule is an electric dipole. The following statement about the water molecule is therefore wrong: (A) The distance between the centres of positive and negative charge in the molecule is a stationary distance. (B) The amount of charge we assign to the positive end and the amount of charge we assign to the negative end of the molecule are equal but opposite. (C) The electric field points from the hydrogen atoms toward the oxygen atom. (D) If we choose the electric potential at infinite distance from the water molecule to be zero, $V = 0$, then the electric potential of the water dipole does not vanish anywhere within a distance of $5 \cdot d$ from the water molecule, where d is the distance between the charged centres of the dipole.

Answer: (D)

Question 13.7
Two point charges repel each other with an electric force of magnitude f_0. If we double both charges, the magnitude of the force between the point charges becomes (A) $f_0/4$, (B) $f_0/2$, (C) f_0 (i.e., it remains unchanged), (D) $2 \cdot f_0$, (E) $4 \cdot f_0$.

Answer: (E)

Question 13.9
Two dipoles at close proximity (A) always repel each other, (B) always attract each other, (C) never interact electrically, (D) attract or repel each other based on their relative orientation.

Answer: (D)

Question 13.11
What have Newton's law of gravity and Coulomb's law in common? (A) force dependence on mass, (B) force dependence on electric charge, (C) force dependence on distance, (D) the magnitude of the proportionality constant, (E) nothing.

Answer: Mathematically, answer (C) is correct; physically it may well be (E) because a unified theory including both forces has not yet been found.

Question 13.13
A mobile positive point charge is located between two parallel charged plates, with the upper plate carrying a positive surface charge density $+\sigma$ and the lower plate carrying a negative surface charge density $-\sigma$. How does the magnitude of the electric field at the position of the mobile point charge change if it moves to twice the distance from the negative plate (i.e., toward the positive plate)? (A) It remains unchanged, (B) it increases by a factor of 4, (C) it doubles, (D) it becomes half of the initial value, (E) it becomes one–quarter of the initial value.

Answer: (A)

Question 13.15
The electric field of a dipole (A) is the same as for a point charge, (B) cannot be described with a formula, (C) decreases faster than for a point charge as we move away from it, (D) decreases slower than for a point charge as we move away from it.

Answer: (C)

Question 13.17

In Millikan's experiment, (A) charged oil droplets are separated by diffusion, (B) neutral oil mist droplets are separated in an electric field, (C) an electric field is applied to an ionic solution, (D) charged oil droplets are levitated in an observation chamber, (E) none of the above.

Answer: (D)

Question 13.19

In which direction does the electric field point at point P for the two charges in Fig. 13.38? Assume that $q_+ = +1\ \mu C$ and $q_- = -1\ \mu C$. *Hint*: Draw the field due to each of the charges separately at P, then see for the resulting direction. (A) left, (B) right, (C) up, (D) down, (E) elsewhere.

Answer: (A)

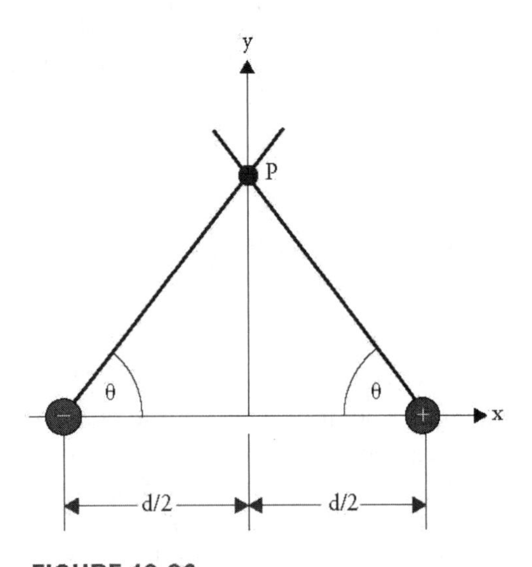

FIGURE 13.38

Question 13.21

In the four processes shown in Fig. 13.39 a mobile point charge (small grey circle with positive or negative charge indicated) is moving in the direction of the arrow shown in the vicinity of an arrangement of stationary charges (either a stationary point charge, depicted as a large grey circle with its charge indicated, or a pair of charged plates). In which case is the work negative (energy is released by the mobile point charge during the displacement)?

Answer: (C)

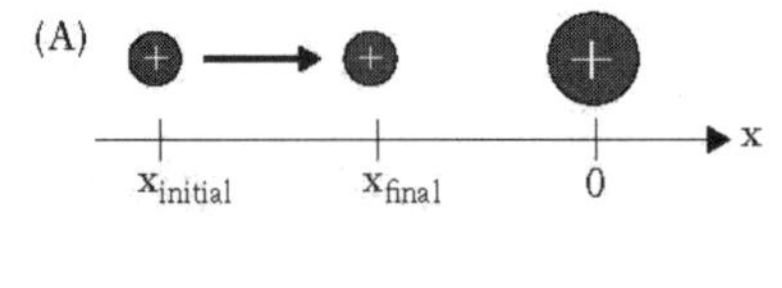

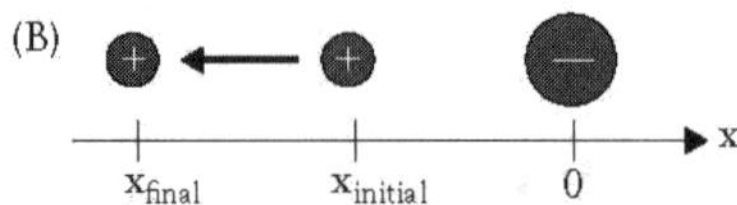

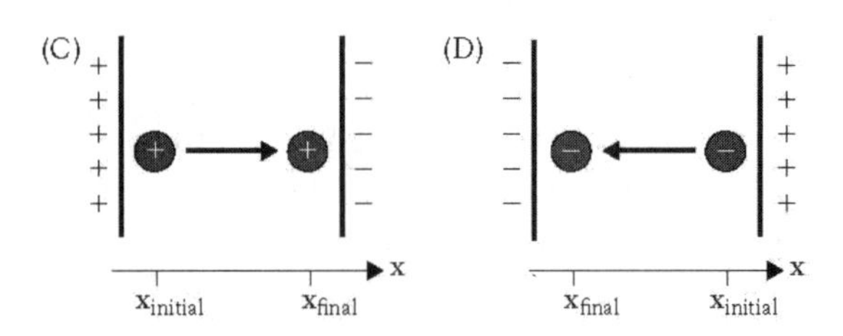

FIGURE 13.39

Question 13.23

Fig. 13.41 shows the electric potential energy of a proton (positive point charge) as a function of distance from a positively charged atomic nucleus, which is located at $r = 0$. Let's assume that the atomic nucleus is very heavy, e.g., the nucleus of a lead atom. We consider two values for the total energy of the proton: proton (1) with total energy $E_{total}(1)$ and proton (2) with total energy $E_{total}(2)$. Which statement about this system is wrong? (A) Neither proton (1) nor proton (2) will travel straight through the nucleus at $r = 0$. (B) Proton (2) will approach the nucleus to closer proximity than proton (1). (C) At the same distance from the nucleus, protons (1) and (2) have the same total energy. (D) At any given position along the r–axis, the electric potential energies for

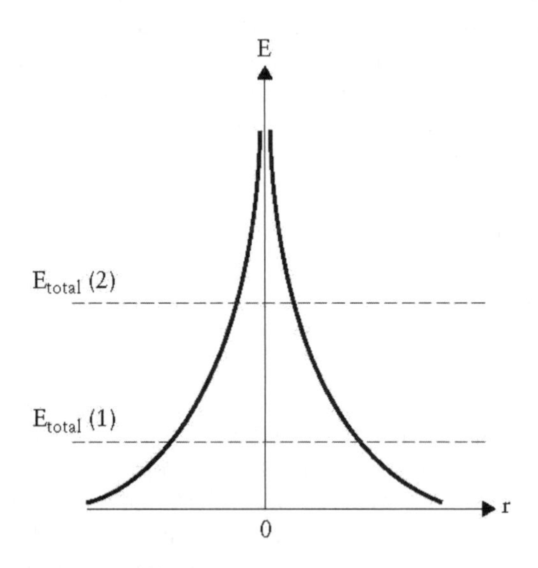

FIGURE 13.41

proton (1) and proton (2) are the same. (E) Proton (1) reaches the same electric potential energy as proton (2) when at infinite distance from the nucleus.

Answer: (C)

Question 13.25

The absolute value of the potential difference between the interior and exterior surfaces of a particular membrane is 65 mV. What additional information will allow you to determine which surface of the membrane carries an excess of positive ions? (A) I need the thickness of the membrane. (B) I need to know which positive ions are involved (e.g., K^+, or Na^+, or Ca^{2+}). (C) I need to establish a Cartesian coordinate system. (D) I need to know the magnitude of the electric field across the membrane. (E) None of the above.

Answer: (E). The direction of the electric field would be sufficient, or identifying the side of the more negative potential.

Question 13.27

The electric potential cannot be given in the following unit: (A) J/C (joule per coulomb), (B) V (volt), (C) N/C (newton per coulomb), (D) (N · m)/C (newton metre per coulomb), (E) More than one of these cannot be used.

Answer: (C)

Question 13.29

If the potential is constant in a certain volume around a given point, what does this mean for the electric field in that volume? (A) The electric field is constant and has a negative value. (B) The electric field is inversely proportional to the distance to the nearest charges. (C) The electric field is zero. (D) The electric field depends linearly on the distance to the nearest charges. (E) The electric field is constant and has a positive value.

Answer: (C)

Question 13.31

If a proton is released from rest in a uniform electric field, does the electric potential at its position increase, stay the same, or decrease?

Answer: It decreases.

ANALYTICAL PROBLEMS

Problem 13.1

We study 3 point charges at the corners of a triangle, as shown in Fig. 13.42. Their charges are $q_1 = 5.0 \times 10^{-9}$ C, $q_2 = -4.0 \times 10^{-9}$ C, and $q_3 = 2.5 \times 10^{-9}$ C. Two distances of separation are also given, $l_{12} = 4$ m and $l_{13} = 6$ m. Find the net electric force on q_3.

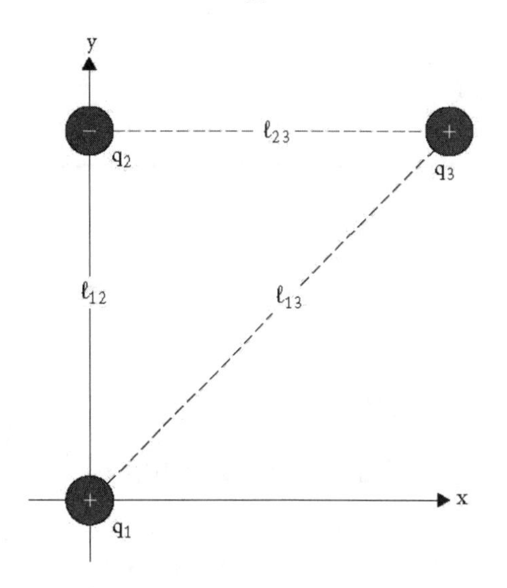

FIGURE 13.42

Solution: The force on point charge q_3 is the superposition of the forces due to point charges q_1 and q_2, i.e., $\mathbf{F}_{13}$ and $\mathbf{F}_{23}$. We calculate each of these two force vectors separately using Coulomb's law. Then we add them to obtain the net force. Starting with the magnitude of the force exerted by q_1 on q_3, $|\mathbf{F}_{13}|$, we find:

$$|\boldsymbol{F}_{13}| = \frac{1}{4 \cdot \pi \cdot \varepsilon_0} \cdot \frac{|q_1|\,|q_3|}{r_{13}^2}$$

$$= \left(9 \times 10^9 \frac{\mathrm{N \cdot m^2}}{\mathrm{C^2}}\right) \frac{(5 \times 10^{-9}\ \mathrm{C})(2.5 \times 10^{-9}\ \mathrm{C})}{(6.0\ \mathrm{m})^2} \quad \textbf{(1)}$$

$$= 3.1 \times 10^{-9}\ \mathrm{N}$$

The direction of this force is along the line connecting point charges q_1 and q_3 and is directed away from point charge q_1 since the two charges repel each other. This leads to a vector which forms an angle θ with the positive x–axis, as illustrated in Fig. 13.51. The angle θ is obtained from geometric analysis of Fig. 13.51. We find, sinθ = 4.0 m/6.0 m = 0.667 which corresponds to θ = 41.8^0. This allows us to express the x– and y–components of the force in the form:

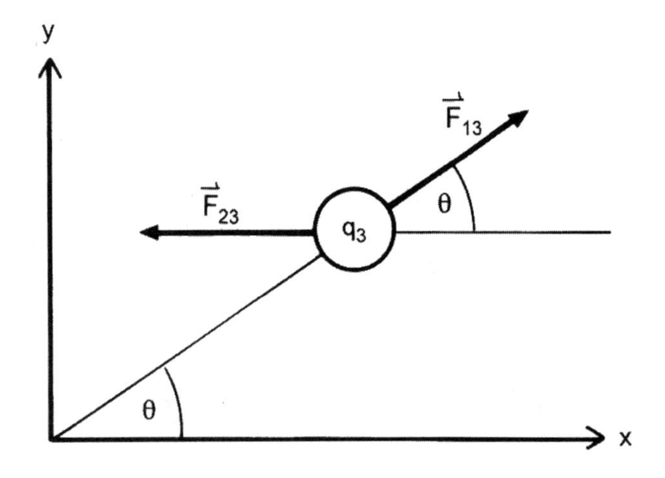

Figure 13.51

$$F_{13,x} = |\boldsymbol{F}_{13}| \cos\theta = 2.31 \times 10^{-9}\,\text{N}$$
$$F_{13,y} = |\boldsymbol{F}_{13}| \sin\theta = 2.07 \times 10^{-9}\,\text{N} \quad \textbf{(2)}$$

We apply again Coulomb's law to obtain the magnitude of the second force, $\mathbf{F}_{23}$. Fig. 13.51 is used for the required distance between the two point charges q_2 and q_3. Applying trigonometry we find:

$$l_{23}^2 + l_{12}^2 = l_{13}^2 \quad \textbf{(3)}$$

which leads to:

$$l_{23} = \sqrt{l_{13}^2 - l_{12}^2} = \sqrt{(6\,\text{m})^2 - (4\,\text{m})^2} = 4.47\,\text{m} \quad \textbf{(4)}$$

Now the given data and the value from Eq. [4] are combined in Coulomb's law:

$$|\boldsymbol{F}_{23}| = \frac{1}{4 \cdot \pi \cdot \varepsilon_0} \cdot \frac{|q_2|\,|q_3|}{r_{23}^2}$$
$$= \left(9 \times 10^9 \frac{\text{N} \cdot \text{m}^2}{\text{C}^2}\right) \frac{(4 \times 10^{-9}\,\text{C})(2.5 \times 10^{-9}\,\text{C})}{(4.47\,\text{m})^2} \quad \textbf{(5)}$$
$$= 4.50 \times 10^{-9}\,\text{N}$$

The direction of this force is along the connecting line between the point charges q_2 and q_3 and is directed toward q_2 since the two charges attract each other. This is illustrated in Fig. 13.51.

With the two individual forces identified, we now calculate the net force acting on q_3. For the x–component we find:

$$F_{net,x} = F_{13,x} - |\boldsymbol{F}_{23}|$$
$$= 2.31 \times 10^{-9}\,\text{N} - 4.50 \times 10^{-9}\,\text{N} \quad \textbf{(6)}$$
$$= -2.19 \times 10^{-9}\,\text{N}$$

for the y–component we get:

$$F_{net,y} = F_{13,y} = +2.07 \times 10^{-9}\,\text{N} \quad \textbf{(7)}$$

Providing the Cartesian components of a vector in Eqs. [6] and [7] is one way to express it. Alternatively, you can calculate the magnitude of the vector, $|\mathbf{F}_{net}|$, and its angle with the positive x–axis, φ (which is the polar coordinate representation). For the vector in Eqs. [6] and [7], this yields:

$$|\boldsymbol{F}_{net}| = \sqrt{F_{net,x}^2 + F_{net,y}^2} = 3.01 \times 10^{-9}\,\text{N} \quad \textbf{(8)}$$

and:

$$\tan\varphi = \frac{F_{net,y}}{F_{net,x}} = \frac{+2.07 \times 10^{-9}\,\text{N}}{-2.19 \times 10^{-9}\,\text{N}} \quad \textbf{(9)}$$

which yields:

$$\varphi = 136.6^0 \quad \textbf{(10)}$$

Problem 13.3
The radius of atomic nuclei follows closely the formula

$$r = 1.2 \times 10^{-15} \cdot A^{1/3} \quad \textbf{(11)}$$

in which r has unit m, and A is the atomic mass in unit g/mol.
(a) Confirm that the density of nuclear matter is independent of the type of atom studied. This density is 2×10^{17} kg/m³!
(b) Using Eq. [11] and A(Bi) = 209.0 g/mol, find the magnitude of the repulsive electrostatic force between two of the protons in a bismuth nucleus when they are separated by the diameter of the nucleus.

<u>Solution to part (a)</u>: Density is defined as $\rho = m/V$. For the material in the nucleus, the mass is the atomic mass,

A. The volume of the nucleus is proportional to the cube of the radius of the nucleus, $V \propto r^3$. Substituting Eq. [11] into this relation, we find for the volume $V \propto (A^{1/3})^3 = A$. Thus, the density is independent of the atomic mass because $\rho = m/V \propto A/A$.

Note that we made an implicit assumption when using the density: $\rho = m/V$ can only be written if the density is constant across the entire volume. Is this assumption justified for a nucleus? Surprisingly, the assumption of a constant density is a very good one for the nuclei of heavier atoms. This is illustrated in Fig. 13.52 showing the charge density in the nucleus. Only close to the edge does the nuclear density tail off.

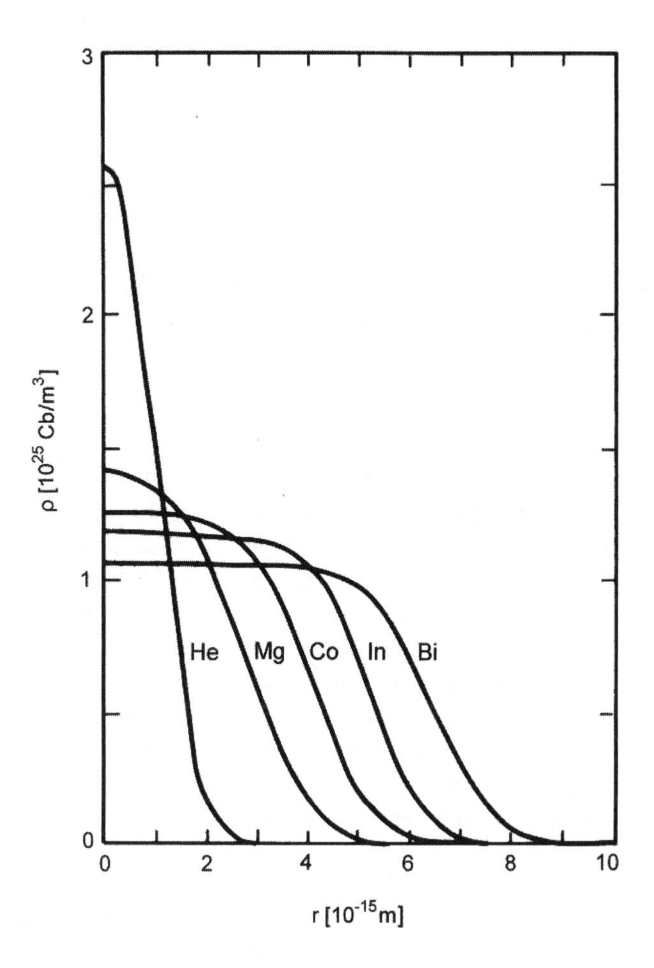

Figure 13.52

Solution to part (b): Substituting the given atomic mass of bismuth in Eq. [11], we find for the radius of the Bi nucleus:

$$r = 1.2 \times 10^{-15} \cdot 209^{1/3} = 7.1 \times 10^{-15}\ \text{m} \quad \textbf{(12)}$$

The diameter of the nucleus is twice its radius. The charge of a proton is equal in magnitude to the elementary charge e, with $e = 1.6 \times 10^{-19}$ C. Therefore, the Coulomb force between two protons equals:

$$|\boldsymbol{F}| = \frac{1}{4 \cdot \pi \cdot \varepsilon_0} \cdot \frac{e^2}{(2 \cdot r)^2}$$

$$= \left(9 \times 10^9 \frac{\text{N} \cdot \text{m}^2}{\text{C}^2}\right) \frac{(1.6 \times 10^{-19}\ \text{C})^2}{(14.2 \times 10^{-15}\ \text{m})^2} \quad \textbf{(13)}$$

$$= 1.1\ \text{N}$$

This is a small force when acting on a macroscopic object, but it is a tremendous force when acting at an atomic length scale. It should lead to the explosion of this and any other nucleus, except for the hydrogen nucleus with only one proton. Luckily, an even stronger attractive force acts between the protons and neutrons in a nucleus, which is the nuclear force.

However, the nuclear force falls off much more steeply than the Coulomb force and the latter starts to dominate at about the distance used in this problem. Therefore, no stable nuclei larger than bismuth's nucleus exist, with two heavier elements only coming close: the thorium isotope Th–232 with a half–life of 14 billion years and the uranium isotope U–238 with a half–life of 4.5 billion years. Note that the latter is not the uranium isotope used in nuclear reactors!

Problem 13.5

A CsCl (cesium chloride) salt crystal is built from the unit cell shown in Fig. 13.44. Cl^- ions form the corners of a cube and a Cs^+ ion is at the centre of the cube. The edge length of the cube, which is called the lattice constant, is 0.4 nm.

(a) What is the magnitude of the net force exerted on the cesium ion by its eight nearest Cl^- neighbours?

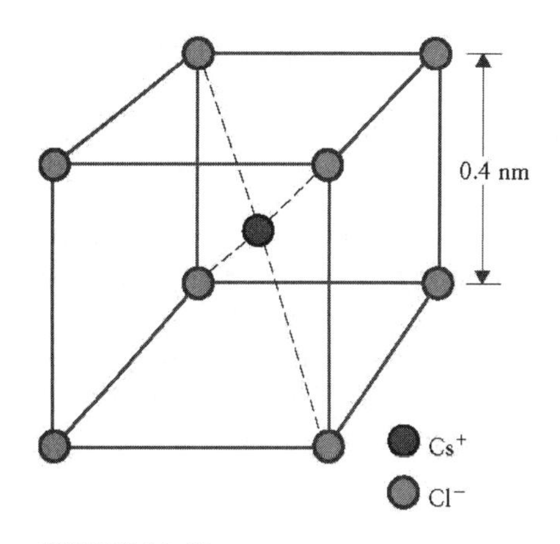

FIGURE 13.44

(b) If the Cl^- in the lower left corner is removed, what is the magnitude of the net force exerted on the cesium ion at the centre by the seven remaining nearest chlorine ions? In what direction does this force act on the cesium ion?

Solution to part (a): Each of the eight Cl^- ions at the corners of the unit cell of the CsCl crystal exerts the same magnitude of electrostatic force on the Cs^+ ion at the centre of the unit cell. All of these pair–wise forces are attractive since they act between a positive Cs ion and a negative Cl ion. Since, therefore, there are four pairs of equal but opposite forces, the net force on the central Cs ion is zero. *Note*: The same is true for the chlorine ions because each Cl^- ion lies in the same fashion at the centre of eight Cs^+ ions. This becomes evident when you keep in mind that the unit cell of the CsCl crystal is repeated in all three Cartesian directions.

Solution to part (b): From an electric point of view, removing a Cl^- ion from one of the corners of the unit cell is the same as adding in Fig. 13.44 a positive elementary charge to that same chlorine ion. Thus, the change from part (a) is one additional force component between that additional positive charge and the central charge of the Cs^+ ion.

We determine first the distance between the added charge and the centre of the unit cell. The distance is half of the length of a line drawn through the cube of side length *a*. From trigonometry we know that the diagonal line in such a cube has the length $\sqrt{3} \cdot a$. Thus, the additional force is:

$$|\boldsymbol{F}| = \frac{1}{4 \cdot \pi \cdot \varepsilon_0} \cdot \frac{e^2}{\left(\frac{\sqrt{3} \cdot a}{2}\right)^2}$$

$$= \left(9 \times 10^9 \frac{\text{N} \cdot \text{m}^2}{\text{C}^2}\right) \frac{(1.6 \times 10^{-19}\ \text{C})^2}{\frac{3}{4}(0.4 \times 10^{-9}\ \text{m})^2} \quad \textbf{(14)}$$

$$= 1.9 \times 10^{-9}\ \text{N}$$

This is a repulsive force, i.e., it acts on the central ion in the direction pointing away from the missing ion at the corner of the unit cell. It is interesting to compare this force with the force we calculated in Problem 13.3, showing the significant difference between electrostatic forces at the length scale of a nucleus and at atomic length scales.

Problem 13.7

Fig. 13.45 shows three positive point charges, with two charges of magnitude q at a distance d along the negative x– and the positive y–axis, and one charge of magnitude $2 \cdot q$ at the origin. Calculate the electric field at point P for $q = 1.0$ nC and a distance $d = 1.0$ m.

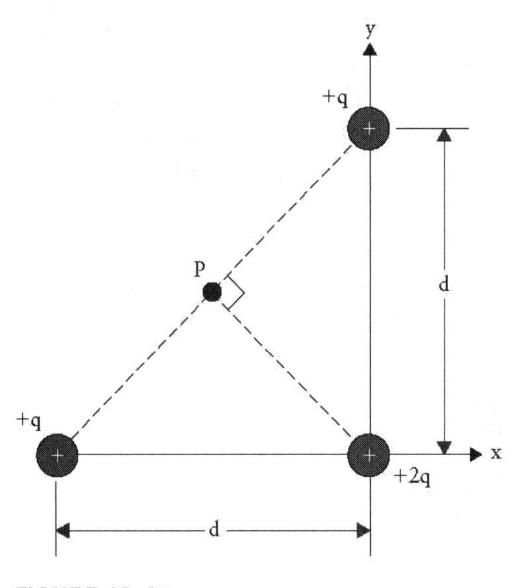

FIGURE 13.45

Solution: The electric field is directly related to the electric force and as such a vector. Thus, for a two–dimensional problem as illustrated in Fig. 13.45, we need to find two electric field components. Fig. 13.45 already identifies the two Cartesian coordinates, the x–direction horizontally to the right and the y–direction vertically up. We are asked to find E_x and E_y.

A calculation like this can become extensive. It is, therefore, highly recommended to search for symmetries in the problem. You will notice for the current problem how this greatly reduces the effort.

We notice that the two charges with value $+q$, at positions $(0, d)$ and $(-d, 0)$, are located at the same distance from point P at opposite sides, i.e., along the dashed line. Thus, the field contributions of these two charges cancel each other. As a result, we need only to determine the field contribution of the charge with value $+2 \cdot q$ to answer the problem.

Since this charge is a point charge, we apply Eq. [13.19] (and not Eq. [13.22], which we would have to choose if more than one point charge would contribute to the net electric field):

$$\boldsymbol{E} = \frac{1}{4 \cdot \pi \cdot \varepsilon_0} \cdot \frac{q_{stationary}}{r^2} \boldsymbol{r}^0 \quad \textbf{(15)}$$

In this equation, the distance from point P to the point charge is required. This distance is determined from Fig. 13.45: point P lies at the centre of a square of side length d. The length of a diagonal line through a square is $\sqrt{2} \cdot d$. Thus, the distance between the point charge and the point P is $\frac{1}{2} \cdot d \cdot \sqrt{2} = d/\sqrt{2}$. This allows us to express the magnitude of the electric field at point P:

$$|\boldsymbol{E}(\boldsymbol{P})| = \frac{1}{4 \cdot \pi \cdot \varepsilon_0} \cdot \frac{2 \cdot q}{\left(\frac{d}{\sqrt{2}}\right)^2} = \frac{q}{\pi \cdot \varepsilon_0 \cdot d^2} \quad \textbf{(16)}$$

The electric field is always directed away from a positive charge. We define an angle θ between the direction of the electric field and the positive x–axis. Thus, with the magnitude in Eq. [16] and the angle θ the task of this problem will be completed. In particular, the two components of the electric field can then be expressed as:

$$E_x = \frac{q}{\pi \cdot \varepsilon_0 \cdot d^2} \cos\theta$$

$$E_y = \frac{q}{\pi \cdot \varepsilon_0 \cdot d^2} \sin\theta \quad \textbf{(17)}$$

The angle θ is obtained from Fig. 13.45: $\theta = 90^0 + 45^0 = 135^0$. Substituting the values given in the problem into Eq. [17] yields for the x–component:

$$E_x = \frac{(1 \times 10^{-9}\ \mathrm{C}) \cos 135^0}{\pi \left(8.85 \times 10^{-12} \frac{\mathrm{C}}{\mathrm{V \cdot m}}\right)(1.0\ \mathrm{m})^2} = -25.4 \frac{\mathrm{V}}{\mathrm{m}} \quad \textbf{(18)}$$

and for the y–component:

$$E_y = \frac{(1 \times 10^{-9}\ \mathrm{C}) \sin 135^0}{\pi \left(8.85 \times 10^{-12} \frac{\mathrm{C}}{\mathrm{V \cdot m}}\right)(1.0\ \mathrm{m})^2} = +25.4 \frac{\mathrm{V}}{\mathrm{m}} \quad \textbf{(19)}$$

Problem 13.9

In Millikan's experiment in Fig. 13.4, a droplet of radius $r = 1.9\ \mu\mathrm{m}$ has an excess charge of two electrons. What are the magnitude and direction of the electric field that is required to levitate the droplet? Use for the density of oil $\rho = 0.925\ \mathrm{g/cm^3}$.

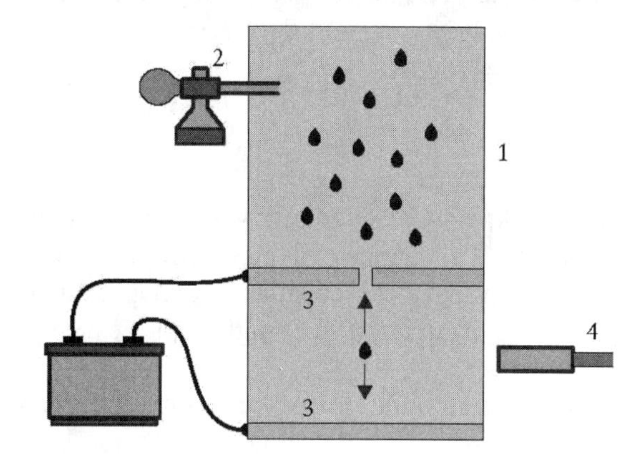

FIGURE 13.4

Solution: To establish a mechanical equilibrium for the droplet, i.e., to maintain a zero–acceleration in the vertical y–direction, an upward acting electrostatic force must compensate the downward directed weight:

$$\sum_i F_{i,y} = 0 = q\,|\boldsymbol{E}| - m \cdot g = q\,|\boldsymbol{E}| - \rho \cdot V \cdot g \quad \textbf{(20)}$$

which results in:

$$n \cdot e\,|\boldsymbol{E}| = \rho\left(\frac{4}{3}\pi \cdot r^3\right) g \quad \textbf{(21)}$$

in which the charge on the oil drop is $q = n \cdot e$, with n an integer number and e the magnitude of the elementary charge. The mass of the spherical droplet has been rewritten as the product of the density and the volume of a sphere of radius r. With $n = 2$, we find for the magnitude of the required electric field:

$$|\boldsymbol{E}| = \frac{4 \cdot \pi \cdot r^3 \cdot \rho \cdot g}{3 \cdot n \cdot e} = \frac{4 \cdot \pi\,(1.9 \times 10^{-6}\ \mathrm{m})^3 \left(925 \frac{\mathrm{kg}}{\mathrm{m^3}}\right)\left(9.8 \frac{\mathrm{m}}{\mathrm{s^2}}\right)}{6\,(1.6 \times 10^{-19}\ \mathrm{C})} = 8.14 \times 10^5 \frac{\mathrm{N}}{\mathrm{C}} \quad \textbf{(22)}$$

While the electrostatic force is directed upward, the electric field is pointing downward with the magnitude calculated in Eq. [22]. This direction of the electric field is due to the negative charge carried by the droplet. To design such a field, a parallel plate capacitor is used with the positively charged plate above the region in which the oil droplet is levitating and the negatively charged plate is below this region.

Problem 13.11
Humid air breaks down electrically when its molecules become ionized. This happens in an electric field $|\mathbf{E}| = 3.0 \times 10^6$ N/C. In that field, calculate the magnitude of the electric force on an ion with a single positive charge.

Solution: This is a straight forward substitution problem for Coulomb's law:

$$F = e\,|\boldsymbol{E}| = (1.6 \times 10^{-19}\ \text{C})\left(3 \times 10^6\ \frac{\text{N}}{\text{C}}\right) = 4.8 \times 10^{-13}\ \text{N} \qquad (23)$$

Problem 13.13
Three positive point charges are located at the corners of a rectangle, as illustrated in Fig. 13.47. Find the electric field at the fourth corner if $q_1 = 3$ nC, $q_2 = 6$ nC, and $q_3 = 5$ nC. The distances are $d_1 = 0.6$ m and $d_2 = 0.2$ m.

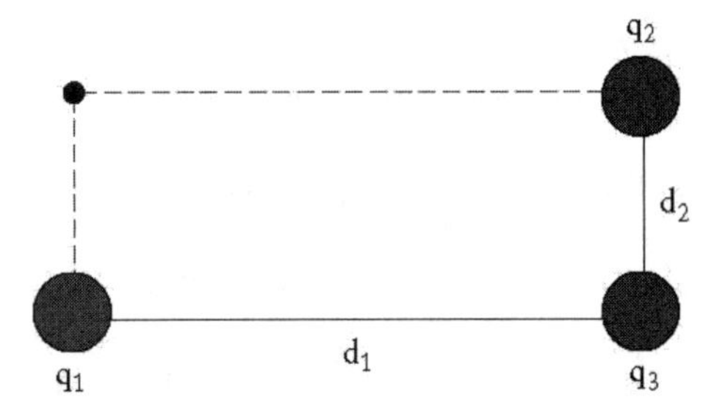

FIGURE 13.47

Solution: The electric field has three contributions. We choose the x–axis to the right and the y–axis upward. The field due to q_1 is (0, 674 N/C), the field caused by q_2 is (– 150 N/C, 0) and the field caused by q_3 is given as (– 106 N/C, + 36 N/C). The net field then results in the form (– 256 N/C, + 710 N/C).

Problem 13.15
(a) What is the electric potential V at a distance $r = 2.1 \times 10^{-8}$ cm from a proton?
(b) What is the electric potential energy in units J and eV of an electron at the given distance from the proton?
(c) If the electron moves closer to the proton, does the electric potential energy increase or decrease?

Solution to part (a): Eq. [13.64] gives the potential for a point charge:

$$V = \frac{1}{4 \cdot \pi \cdot \varepsilon_0} \cdot \frac{e}{r} = \left(9 \times 10^9\ \frac{\text{N} \cdot \text{m}^2}{\text{C}^2}\right) \frac{1.6 \times 10^{-19}\ \text{C}}{2.1 \times 10^{-10}\ \text{m}} = 6.86\ \text{V} \qquad (24)$$

Solution to part (b): The electric potential energy follows from the charge and the potential:

$$E_{el} = q \cdot V = (-1.6 \times 10^{-19}\ \text{C})\,(6.86\ \text{V}) = -1.1 \times 10^{-18}\ \text{J} \qquad (25)$$

in which the negative sign is due to the opposite signs of the two interacting charges. This means that the electron is attracted by the proton.

Solution to part (c): As the electron comes closer to the proton the electric potential energy becomes even more negative than calculated in Eq. [25]. Since the electron and proton attract each other, this is the same situation as we saw before for gravity, where two masses always attract each other and the gravitational potential energy is lowered when the masses approach each other (e.g., when objects fall toward the surface of Earth). In a classical sense, the electron does not crash into the proton for the same reason the Moon doesn't fall down onto Earth: the circular motion (Moon around Earth or electron around proton) leads to an apparent centrifugal effect balancing the attractive force.

Problem 13.17
We study the three point charges shown in Fig. 13.49. They are held at the corners of an equilateral triangle with $l = 0.2$ m. What is the electric potential energy of the system of three point charges? Use for the 3 charges: $q_1 = +2\,Q$, $q_2 = -3\,Q$ and $q_3 = +Q$ where $Q = 100$ nC. *Hint*: The solution is done in steps. Assume that you first bring one of the point charges from a very large (infinite) distance to its position. Then repeat this procedure for the second and third point charges.

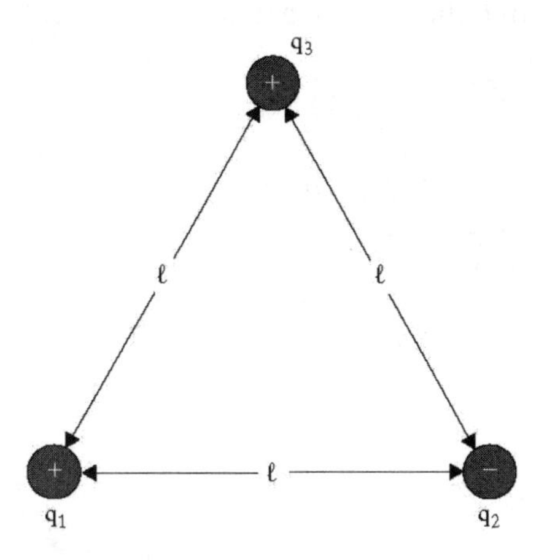

FIGURE 13.49

Solution: Following the hint, we built the system of charges by moving one after the other from a place where they are all at infinite distance to each other. Since the electric force decreases rapidly with distance, a distance equivalent to an infinite distance might practically be a rather short distance. In vacuum this distance is quite large for a precise measurement; however, in a solution a few micrometres is an infinite distance due to charge screening effects. The electric energy of the system is $E_{el} = 0$ J while the charges are at infinite distance to each other, i.e., while they do not interact with each other.

When we bring q_1 to its final position, the electric energy remains zero, i.e., $E_{el} = 0$ J, since the charge still does not interact with any other charge. Next we bring q_2 to its final position, which is at a distance d from charge q_1. In this process the change in the electric potential energy is:

$$\Delta E_{el,\,12} = \frac{1}{4 \cdot \pi \cdot \varepsilon_0} \cdot \frac{q_1 \cdot q_2}{d} \quad \textbf{(26)}$$

In the last step we bring q_3 to its final position at the distance d to either of the other two point charges. The work associated with this step is equal to the sum of the work to bring q_3 to a distance d from q_1 *and* the work to bring q_3 to a distance d from q_2:

$$W_{13} + W_{23} = \Delta E_{el,\,13} + \Delta E_{el,\,23} = \frac{1}{4 \cdot \pi \cdot \varepsilon_0} \cdot \frac{q_1 \cdot q_3}{d} + \frac{1}{4 \cdot \pi \cdot \varepsilon_0} \cdot \frac{q_2 \cdot q_3}{d} \quad \textbf{(27)}$$

The total change of the electric potential energy is the sum of the contributions due to each of the three steps toward the final charge arrangement:

$$E_{el} = \Delta E_{el,\,12} + \Delta E_{el,\,13} + \Delta E_{el,\,23} = \frac{1}{4 \pi \varepsilon_0} \left(\frac{2\,Q(-3\,Q)}{d} + \frac{2\,Q \cdot Q}{d} + \frac{(-3\,Q)\,Q}{d} \right) \quad \textbf{(28)}$$

This yields:

$$E_{el} = -\frac{7 \cdot Q^2}{4 \cdot \pi \cdot \varepsilon_0 \cdot d} \quad \textbf{(29)}$$

substituting the given values leads to the final result:

$$E_{el} = \left(-9 \times 10^9 \, \frac{\text{N m}^2}{\text{C}^2} \right) \frac{7 \cdot (100 \times 10^{-9}\ \text{C})^2}{0.2\ \text{m}} = -3.2 \times 10^{-3}\ \text{J} \quad \textbf{(30)}$$

Since the work calculated in Eq. [30] is negative, we know that the source of an external force has to do this amount of work to *disassemble* the structure of charges as shown in Fig. 13.49. In turn, creating the arrangement of charges shown in Fig. 13.49 by starting with the three charges infinitely separated releases 3.2 mJ of energy.

Problem 13.19

Fig. 13.50 shows three positive point charges at the corners of a rectangle. Find the electric potential at the upper right corner if $q_1 = 8$ μC, $q_2 = 2$ μC, and $q_3 = 4$ μC. The distances are $d_1 = 6.0$ cm and $d_2 = 3.0$ cm. The potential is defined such that it is 0 V at infinite distance from the point charge arrangement shown.

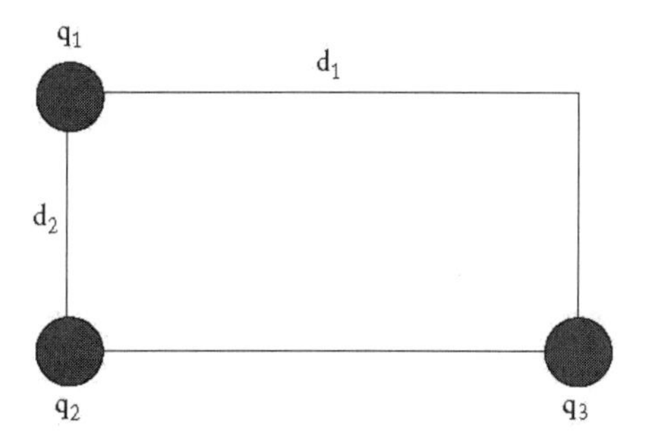

FIGURE 13.50

Solution: The electric potential at the upper right corner is the sum of the potentials created by each of the three charges. To find the potential created by each charge we need both the charge itself (given in the problem) and the distance between each charge and the corner (from the geometry of the rectangle). The distance between charge q_1 and the corner is $r_1 = d_1 = 0.060$ m. The distance $r_3 = d_2 = 0.030$ m. The distance r_2 is:

$$r_2 = \sqrt{d_1^2 + d_2^2} = \sqrt{(0.060\text{ m})^2 + (0.030\text{ m})^2} = 0.067\text{ m} \quad \textbf{(31)}$$

Now we can find the total potential:

$$V = V_1 + V_2 + V_3 = \frac{1}{4\pi\varepsilon_0}\left(\frac{q_1}{r_1} + \frac{q_2}{r_2} + \frac{q_3}{r_3}\right) = 2.7 \times 10^6\ \frac{\text{N}}{\text{C}} \quad \textbf{(32)}$$

CHAPTER FOURTEEN

Nervous system: The flow of charges

MULTIPLE CHOICE AND CONCEPTUAL QUESTIONS

Question 14.1
How does the capacitance of a parallel plate capacitor change when its plates are moved to twice their initial distance and a slab of material with dielectric constant $\kappa = 2$ is placed between the plates to replace air? (A) It is increased to 4 times the original value. (B) It doubles. (C) It remains unchanged. (D) It is halved. (E) It is reduced to one–quarter of the original value.

Answer: (C)

Question 14.3
A capacitor provides an electric field that points from left to right. How will the water molecule shown in Fig. 13.1 rotate if it can rotate freely like in liquid water? (Assume its orientation as shown; i.e., left in the context of this question is left on the page.) (A) It will not rotate. (B) It will rotate such that the oxygen atom is left and the two hydrogen atoms are right. (C) It will rotate such that the oxygen atom is right and the two hydrogen atoms are left. (D) It will rotate such that the oxygen atom is up and the two hydrogen atoms are down. (E) It will rotate such that the oxygen atom is down and the two hydrogen atoms are up.

Answer: (B)

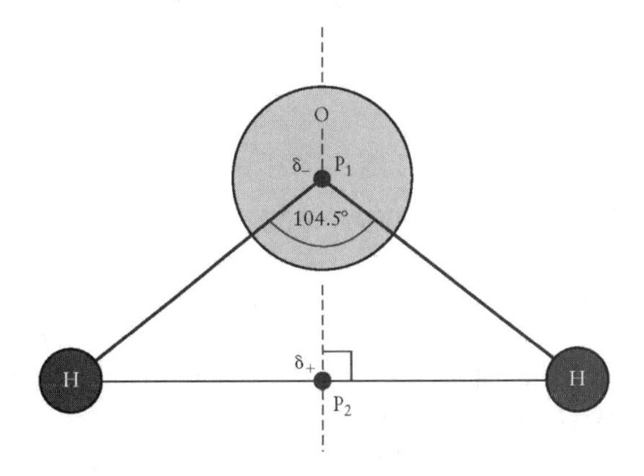

FIGURE 13.1

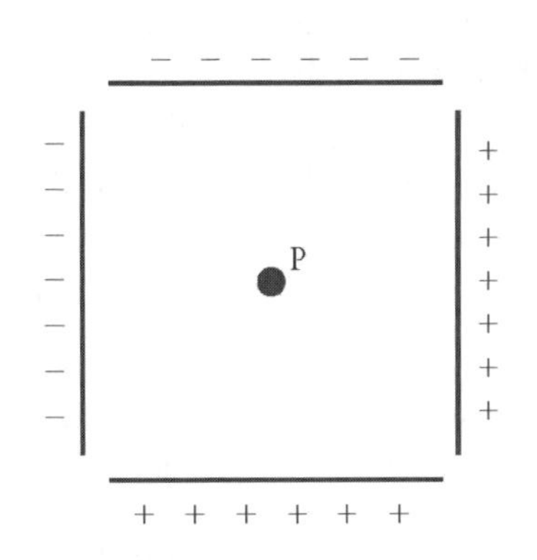

FIGURE 14.31

Question 14.5
Fig. 14.31 shows two capacitors that are rotated 90^0 relative to each other. In what direction does the electric field point at point P at the centre of the arrangement? (A) up, (B) down, (C) left, (D) right, (E) in another direction.

Answer: (E), to the upper left.

Question 14.7
How does the capacitance of an air–filled parallel plate capacitor change when we place a material with dielectric constant $\kappa = 1.0$ between its plates? (Choose the closest value.) (A) It triples. (B) It doubles. (C) It remains unchanged. (D) It is halved. (E) It is reduced to one–third of the original value.

Answer: (C)

Question 14.9
We compare myelinated and unmyelinated nerves. Remember that the latter have a much smaller membrane thickness. What consequence does this have for human nerves, which all have a resting potential difference of – 70 mV? (A) The capacitance per unit membrane area of the myelinated nerve is larger than the capacitance per unit membrane area of the unmyelinated nerve. (B) The dielectric constant is much smaller for the myelinated nerves. (C) The potential difference across the myelinated membrane must be larger. (D) The amount of charge separated across the membrane per unit area is larger for the unmyelinated nerve. (E) The amount of charge separated across the membrane per unit area is larger for the myelinated nerve.

Answer: (D)

Question 14.11

Why is it dangerous to touch the terminals of a capacitor with a high potential difference even after the charging source has been disconnected? What would you do to make the capacitor safer to handle?

Answer: The plates remain charged when the battery is disconnected. When you touch it, the capacitor can discharge through your body (like a defibrillator does when applied to the chest of a patient). Prior to handling the device you have to connect all capacitors to ground potential.

Question 14.13

If the potential difference across a capacitor is doubled, by what factor does the stored electric energy change?

Answer: It will increase by a factor of four due to Eq. [14.8]:

$$\Delta E_{el} = \frac{1}{2} C (\Delta V_{final})^2 \quad (1)$$

Note, you cannot answer by using Eq. [14.7],

$$\Delta E_{el} = \frac{1}{2} Q \cdot \Delta V_{final} \quad (2)$$

because the potential change is achieved by changing the amount of charges on the plates. See also Question 14.14.

Question 14.15

Explain why all points in a conductor must have the same potential under stationary conditions.

Answer: Potential differences cause the flow of charges in a conductor. This would cause non–stationary conditions.

Question 14.17

(a) Why don't free electrons in a piece of metal fall to the bottom due to gravity?
(b) Why don't free electrons in a piece of metal all drift to the surface due to mutual electric repulsion based on Coulomb's law?

Answer to part (a): Because their electric interaction is many orders of magnitude greater than the gravitational pull on an electron. Thus, gravity effects can be neglected and the electric electron/core ion interaction keeps them uniformly distributed.

Answer to part (b): Because the immobile core ion to which the respective electron belongs carries a positive charge. Electrons cannot move far from their respective core ion unless another electron moves into the vicinity of the core ion, as happens when a current flows. Note that *excess* electrons brought into the metal do indeed drift to the surface as stated in the question.

Question 14.19

Which property causes point charges to move along a conductor? (A) The conductor's resistance, (B) the conductor's resistivity, (C) the electric current, (D) the electric potential, or (E) the electric charge of the electron.

Answer: (D)

Question 14.21

Which is the standard unit for resistance? (A) Ω, (B) Ω/m², (C) Ω · m², (D) Ω/m, (E) Ω · m.

Answer: (A)

Question 14.23

Which is the standard unit for resistivity? (A) V/A, (B) A/V, (C) none; it is a unitless materials constant, (D) same as for resistance, (E) none of the above.

Answer: (E). It differs from the unit of resistance by an additional unit of metre (m).

Question 14.25

Which is the standard unit of the drift velocity? (A) V/Ω², (B) V/A, (C) Ω/m, (D) m/s, (E) V²/C.

Answer: (D)

Question 14.27

The electric system in Fig. 14.32 is often used to model a nerve membrane. It consists of a series of identical resistors aligned in two parallel rows with identical capacitors bridging the two rows following each resistor. Which of the following assumptions in order (A) to (D) has to be revised for Fig. 14.32 to be a reasonable model for a nerve (rendering the subsequent assumptions invalid as well)? (A) The membrane can be divided into segments. (B) The segments can be chosen such that they have the same capacitance. (C) The resistances in each row are the same. (D) For each single segment, the resistances on both sides of the membrane are the same. (E) None of these assumptions needs to be revised.

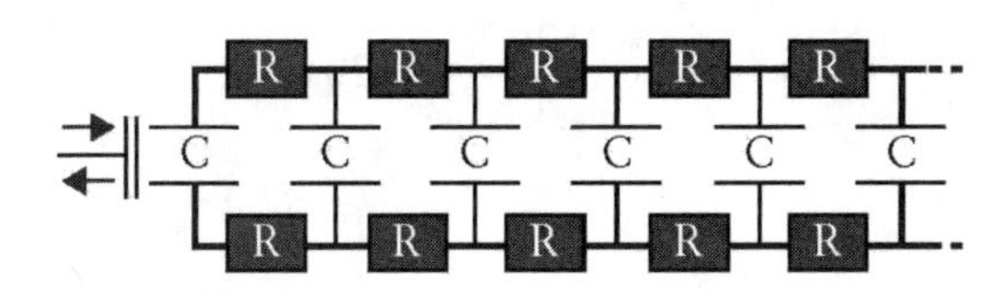

FIGURE 14.32

Answer: (D). Axoplasm and extracellular fluid do not necessarily have the same resistivity and geometry for current flow.

ANALYTICAL PROBLEMS

Problem 14.1
We study some capacitor arrangements.
(a) An air–filled parallel plate capacitor has a plate separation of $b = 1.5$ mm and an area $A = 4.0\ \text{cm}^2$. Find its capacitance.
(b) A capacitor with capacitance of $C = 4.5\ \mu\text{F}$ is connected to a 9-V battery. What is the amount of charge on each plate of the capacitor?

Solution to part (a): The capacitance is given in Eq. [14.3]:

$$C = \varepsilon_0 \frac{A}{b} = \left(8.85 \times 10^{-12} \frac{\text{C}^2}{\text{N m}^2}\right) \frac{4 \times 10^{-4}\ \text{m}^2}{1.5 \times 10^{-3}\ \text{m}} \quad (3)$$

$$= 2.36 \times 10^{-12}\ \text{F} = 2.36\ \text{pF}$$

Solution to part (b): The charge follows from Eq. [14.2]:

$$q = C \cdot \Delta V = (4.5 \times 10^{-6}\ \text{F})(9\ \text{V})$$
$$= 4.05 \times 10^{-5}\ \text{C} = 40.5\ \mu\text{C} \quad (4)$$

Problem 14.3
An air–filled parallel plate capacitor has a capacitance of 60 pF.
(a) What is the separation of the plates if each plate has an area of $0.5\ \text{m}^2$?
(b) If the region between the plates is filled with a material with $\kappa = 4.5$, what is the final capacitance?

Solution to part (a): A capacitor with a vacuum between its conducting plates is rare. For any other material between the capacitor plates, the formula for the capacitance in Eq. [3] has to be replaced by

$$C_{dielectric} = \frac{\kappa_{dielectric} \cdot \varepsilon_0 \cdot A}{b} = \kappa_{dielectric} \cdot C_{vacuum} \quad (5)$$

If the capacitor is air–filled, however, it is justified to make the approximation $\kappa = 1$ as Table 14.2 suggests. Using Eq. [3], we solve for the gap spacing of the capacitor, b:

$$b = \varepsilon_0 \frac{A}{C} = \left(8.85 \times 10^{-12} \frac{\text{C}^2}{\text{N m}^2}\right) \frac{0.5\ \text{m}^2}{6 \times 10^{-11}\ \text{F}} \quad (6)$$

$$= 7.4 \times 10^{-2}\ \text{m} = 7.4\ \text{cm}$$

Table 14.2

Material	Dielectric constant κ
Vacuum	1.0
Air at 1.0 atm	1.00054
Polystyrene	2.6
Paper	3.5
Pyrex glass	4.7
Porcelain	6.5
Nerve membrane	7.0
Silicon	12.0
Ethanol	25.0
Water	78.5

Solution to part (b): Instead of rewriting the same formulas, we choose a relative approach to express the result. From Eq. [5] we know that $C \propto \kappa$, thus:

$$C_{material} = \frac{\kappa_{material}}{\kappa_{air}} C_{air}$$
$$= \frac{4.5}{1.0} 6 \times 10^{-11}\ \text{F} = 2.7 \times 10^{-10}\ \text{F} = 270\ \text{pF} \quad (7)$$

Problem 14.5
An air–filled parallel plate capacitor has a plate separation of 0.1 mm. What plate area is required to provide a

capacitance of 2.0 pF?

Solution: Since we have the capacitance and the plate separation, we can re–arrange the capacitance equation [14.3] to solve for the area. Don't forget that capacitance should be in farads, not picofarads and the plate separation in metres: $A = 2.3 \times 10^{-5}\ m^2 = 0.23\ cm^2$.

Problem 14.7
An air–filled parallel plate capacitor has a plate area of 2.0 cm² and plate separation of 5.0 mm. If a 12.0-V battery is connected to its plates, how much energy does the device store?

Solution: The energy stored in a capacitor can be found using any two of the potential difference between the plates, the capacitance and the charged stored on the capacitor. We start by finding the capacitance $C = 3.5 \times 10^{-13}$ F.

Since we also know the potential difference it is easiest to use Eq. [1] to find $E_{el} = 2.55 \times 10^{-11}$ J.

Problem 14.9
A parallel plate capacitor carries a charge Q on plates of area A. A dielectric material with dielectric constant κ is located between its plates. We can show that the force each plate exerts on the other is given by:

$$F = \frac{Q^2}{2 \cdot \kappa \cdot \varepsilon_0 \cdot A} \quad \textbf{(8)}$$

When a potential difference of 0.1 kV exists between the plates of an air–filled parallel plate capacitor of C = 20 μF capacitance, what force do the two plates exert on each other if they are separated by 2.0 mm?

Solution: In order to find the force exerted on each plate using the given equation we need to find the charge stored in the capacitor using $Q = C \cdot \Delta V = 2.0 \times 10^{-3}$ C.

We can then make a substitution in the given equation by re–arranging the equation for the capacitance in the form:

$$C \cdot b = \kappa \cdot \varepsilon_0 \cdot A \quad \textbf{(9)}$$

which we can substitute into the given force equation to find: $F = 50$ N.

Problem 14.11
All commercial electric devices have identifying plates that specify their electrical characteristics. For example, a typical household device may be specified for a current of 6.0 A when connected to a 120-V source. What is the resistance of this device?

Solution: We use Ohm's law, in the form of Eq. [14.28]:

$$R = \frac{\Delta V}{I} = \frac{120\ \text{V}}{6\ \text{A}} = 20\ \Omega \quad \textbf{(10)}$$

Problem 14.13
A rectangular piece of copper is 2 cm long, 2 cm wide and 10 cm deep.
(a) What is the resistance of the copper piece as measured between the two square ends? (Use the resistivity of copper from Table 14.3.)
(b) What is the resistance between two opposite rectangular faces?

Table 14.3

Material	Resistivity (Ω · m)
Insulators and semiconductors:	
Yellow sulfur	2.0×10^{15}
Artificial lipid membrane	1.0×10^{13}
Quartz	1.0×10^{13}
Nerve membrane	1.6×10^{7}
Silicon	2.5×10^{3}
Axoplasm	1.1×10^{0}
Germanium	5.0×10^{-1}
Metals:	
Mercury	1.0×10^{-6}
Iron	1.0×10^{-7}
Gold	2.4×10^{-8}
Copper	1.7×10^{-8}

Solution to part (a): We use Eq. [14.29] to relate the resistance and resistivity:

$$R = \frac{\rho \cdot l}{A} \quad \textbf{(11)}$$

Each square end of the copper block has an area of $A_a = (2 \times 10^{-2}\ m)^2 = 4 \times 10^{-4}\ m^2$. With a length of the block of $l_a = 0.1$ m, we get:

$$R = \frac{(1.7 \times 10^{-8}\ \Omega\ \text{m})\ (0.1\ \text{m})}{4 \times 10^{-4}\ \text{m}^2} = 4.25 \times 10^{-6}\ \Omega = 4.25\ \mu\Omega \quad \textbf{(12)}$$

Solution to part (b): All 4 rectangular faces of the copper block have the same area, $A_b = 2 \times 10^{-2}\,m \cdot 0.1\,m = 2 \times 10^{-3}\,m^2$. Using Eq. [11] with the length of the block now $l_b = 2 \times 10^{-2}$ m we find $R = 1.7 \times 10^{-7}\,\Omega = 0.17\,\mu\Omega$. Note that the difference between the resistances in parts (a) and (b) is entirely due to the change in geometry.

Problem 14.15

A conducting, cylindrical wire has a diameter of 1.0 mm, a length of 1.67 m and a resistance of 50 mΩ. What is the resistivity of the material? Identify the material of which this conductor is made by using Table 14.3.

Solution: We use Eq. [11] to determine the resistivity of the wire material. Its cross–sectional area is $A = r^2 \cdot \pi = \pi\,(0.5 \times 10^{-3}\,m)^2 = 7.85 \times 10^{-7}\,m^2$, where we converted the given diameter to a radius. Thus:

$$\rho = \frac{R \cdot A}{l} = \frac{(50 \times 10^{-3}\,\Omega)\,(7.85 \times 10^{-7} m^2)}{1.67\ m} = 2.35 \times 10^{-8}\,\Omega \cdot m \quad (13)$$

Comparing with the tabulated values of resistivities, this value comes closest to the value of gold (Au).

Problem 14.17

You often see birds resting on power lines that carry currents of 50 A. The copper wire on which the bird stands has a radius of 1.1 cm. Assuming that the bird's feet are 4.0 cm apart, calculate the potential difference across its body.

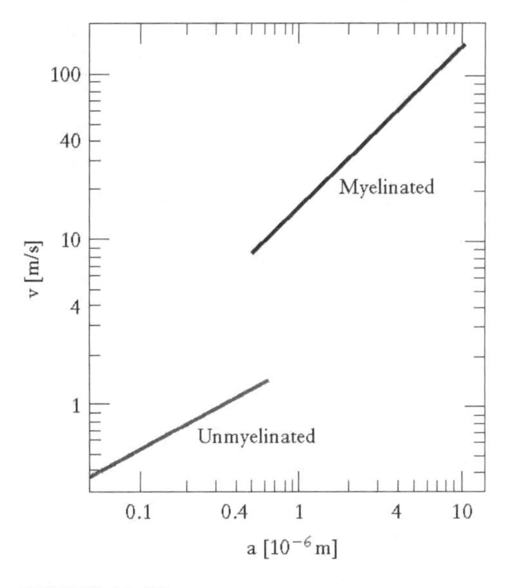

FIGURE 14.29

Solution: We need to look at the potential difference between two points in a copper wire of cross–sectional area $A = \pi \cdot (1.1 \times 10^{-2}\,m)^2 = 3.8 \times 10^{-4}\,m^2$ and the length $l = 0.04$ m. Using Ohm's law we find $\Delta V = 89\,\mu V$.

Problem 14.19

A current density of $0.8 \times 10^{-4}\,A/cm^2$ stimulates a 6-nm-thick nerve membrane for 150 μs. How does the potential across the membrane change as a result of this current density?

Solution: The current across the membrane, I_m, and the time the current flows combine to the charge that is separated across the membrane. This charge is used in the definition of the capacitance to calculate the potential difference:

$$\Delta V = \frac{|I_m|\,\Delta t}{C} = \frac{J_m \cdot A \cdot \Delta t}{C} \quad (14)$$

In Eq. [14] the current is replaced by the current density because this quantity is given in the problem text. Next we substitute the capacitance from Eq. [5] and enter the numerical values, including the dielectric constant of the nerve membrane from Table 14.2 and the current density $J_m = 0.8 \times 10^{-4}\,A/cm^2$, which is the same as $0.8\,A/m^2$:

$$\Delta V = \frac{J_m \cdot A \cdot \Delta t}{\kappa \cdot \varepsilon_0 \dfrac{A}{b}} = \frac{J_m \cdot b \cdot \Delta t}{\kappa \cdot \varepsilon_0} = \frac{\left(0.8\,\dfrac{A}{m^2}\right)(6 \times 10^{-9}\,m)\,(150 \times 10^{-6}\,s)}{7.0\left(8.85 \times 10^{-12}\,\dfrac{C^2}{N \cdot m^2}\right)} = 0.012\ V = 12\ mV \quad (15)$$

Problem 14.21

Confirm both relations in Eq. [14.83] graphically by using Fig. 14.29:

$$\begin{aligned} &\textit{unmyelinated nerve:} \quad v \propto a^{0.5} = \sqrt{a} \\ &\textit{myelinated nerve:} \quad v \propto a^{1.0} = a \end{aligned} \quad (16)$$

Solution: We start with the unmyelinated nerves. Fig.

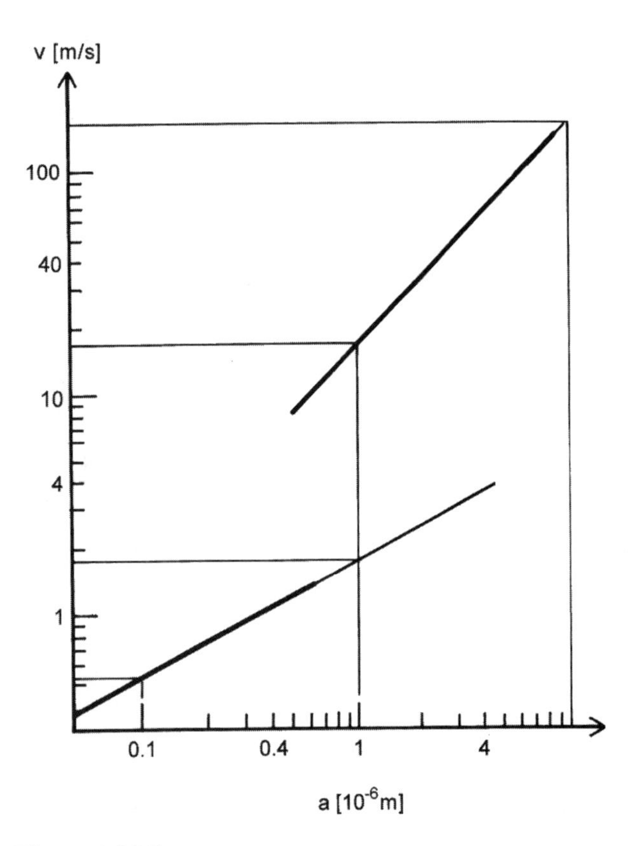

Figure 14.35

Table 14.6

a (μm)	lna	v (m/s)	lnv
0.1	– 2.303	0.51	– 0.67
1.0	0.0	1.75	+ 0.56

Solution: We start with the unmyelinated nerves. Fig. 14.35 shows a particular choice of two points in the double–logarithmic graph of Fig. 14.29 for the data analysis. The data read off the figure are shown in Table 14.6. We write the general power law relation between the radius of the axon and the impulse speed (with speed v given in unit m/s and parameter a given in unit μm):

$$v = c_2 \cdot a^{c_1} \quad \textbf{(17)}$$

with c_1 and c_2 constants. This yields:

$$\ln v = \ln c_2 + c_1 \cdot \ln a \quad \textbf{(18)}$$

In this formula we need only to evaluate c_1 because Eq. [16] does predict a power law coefficient, but no prefactor. For the data in Table 14.6 we find:

$$c_1 = \frac{\Delta \ln v}{\Delta \ln a} = \frac{1.23}{2.303} = 0.53 \quad \textbf{(19)}$$

Next we analyse the data for the myelinated nerve in Fig. 14.29. Fig. 14.35 shows again a choice of data points to quantify the power–law behaviour. The data are listed in Table 14.7.

Table 14.7

a (μm)	lna	v (m/s)	lnv
1.0	0.0	16.1	+ 2.78
10.0	+ 2.303	160.0	+ 5.08

Using again Eq. [18], we find for the power law coefficient for myelinated nerves:

$$c_1 = \frac{\Delta \ln v}{\Delta \ln a} = \frac{2.3}{2.303} = 1.0 \quad \textbf{(20)}$$

Both exponents are consistent with Eq. [16].

Problem 14.23

A simplified model for an erythrocyte is a spherical capacitor with a positively charged liquid interior of surface area A. The interior fluid is separated by a membrane of thickness b from the surrounding, negatively charged plasma fluid. The potential difference across the membrane is 100 mV and the thickness of the membrane is about 100 nm with a dielectric constant of $\kappa = 5.0$.
(a) Calculate the volume of the blood cell assuming that an average erythrocyte has a mass 1×10^{-12} kg. From the volume determine the surface area of the erythrocyte.
(b) Calculate the capacitance of the blood cell. For this calculation, model the membrane as a parallel plate capacitor with the area found in part (a).
(c) Calculate the charge on the surface of the membrane. How many elementary charges does this represent? Use 1.06 g/cm³ as the density of blood.

Solution to part (a): The first part of the question serves as a preparation step to obtain some necessary parameters to quantify later the electric properties.

The volume of the blood cell is estimated from the mass of the blood cell and the density of blood. The density of blood is a sufficiently good approximation of the density of the blood cell as blood cells float in whole blood. We find for the volume:

$$V = \frac{1 \times 10^{-12}\ \text{kg}}{1.06 \times 10^{3}\ \frac{\text{kg}}{\text{m}^3}} = 9.4 \times 10^{-16}\ \text{m}^3 \tag{21}$$

The volume of a spherical blood cell is $V = 4 \cdot \pi \cdot r^3/3$ and its surface is $A = 4 \cdot \pi \cdot r^2$. From Eq. [21], we obtain $r = 6.1 \times 10^{-6}$ m = 6.1 μm. This yields:

$$A = 4.7 \times 10^{-10}\ \text{m}^2 \tag{22}$$

Solution to part (b): Treating the membrane as a parallel plate capacitor we find from Eq. [5]:

$$C = 5.0\left(8.85 \times 10^{-12}\ \frac{\text{F}}{\text{m}}\right)\frac{4.7 \times 10^{-10}\ \text{m}^2}{100 \times 10^{-9}\ \text{m}} = 2.08 \times 10^{-13}\ \text{F} \tag{23}$$

Solution to part (c): Using $q = C \cdot \Delta V$ we find from part (b) for the charge:

$$q = (2.08 \times 10^{-13}\ \text{F})(100 \times 10^{-3}\ \text{V}) = 2.08 \times 10^{-14}\ \text{C} \tag{24}$$

Dividing this charge by the elementary charge e allows us to calculate the number of elementary charges on the surface membrane of the blood cell:

$$N = \frac{q}{e} = \frac{2.08 \times 10^{-14}\ \text{C}}{1.6 \times 10^{-19}\ \text{C}} = 130000 \tag{25}$$

CHAPTER FIFTEEN

Electrocardiography: Electric phenomena of the heart

There are no multiple choice and conceptual questions, and no analytical problems with this chapter.

CHAPTER SIXTEEN

Elastic tissue: Elasticity and vibrations

MULTIPLE CHOICE AND CONCEPTUAL QUESTIONS

Question 16.1

The dash–dotted curve in Fig. 16.2(a) is the active stretching force for a human muscle. This curve shows the force (as a fraction of the maximum force in %) a muscle can exert at a given length. For example: the maximum force can be exerted when the muscle is at its resting length, but only about 50% of that force can be exerted when the muscle has contracted to 80% of its resting length. The dashed curve in Fig. 16.2(a) shows the passive stretching force for a human muscle. This curve indicates the external force required to stretch the muscle at the given length. For example: to stretch a muscle when it is about 5% longer than its resting length requires about double the force as stretching the same muscle just beyond its resting length. Which of the following statements is true? (A) Hooke's law is suitable to describe the passive stretching of a muscle near its resting length. (B) Hooke's law is suitable to describe the active muscle force for a given muscle near its resting length. (C) Fig. 16.2(a) does not allow us to verify statements like those made in (A) and (B). (D) Both curves in Fig. 16.2(a) indicate that muscle tissue has no elastic regime near the resting length. (E) Fig. 16.2(a) implies that a muscle released after stretching should undergo simple harmonic motion.

Answer: (D)

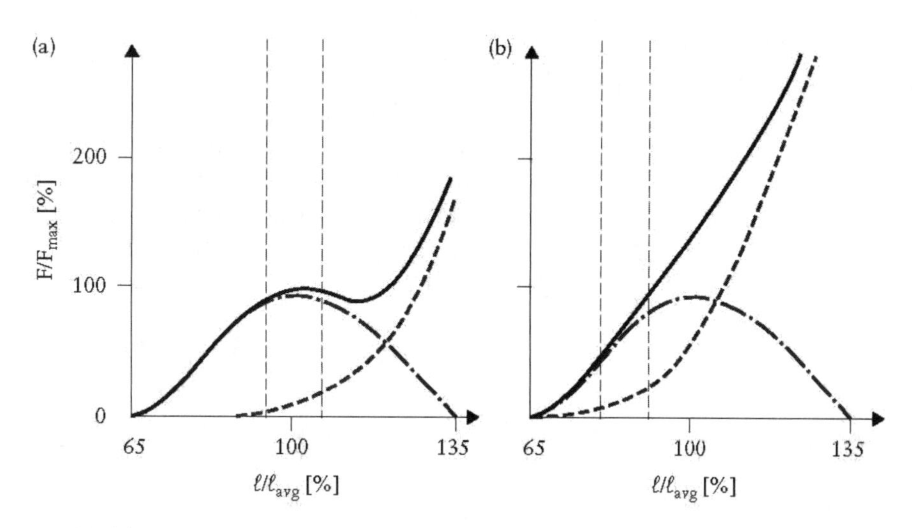

FIGURE 16.2

Question 16.3

The elastic behaviour of the blood vessel in Fig. 16.10 is the result of two contributing components in the tissue: elastin (dash–dotted curve) and collagen, with a bulk modulus increasing with strain (dashed curve). Based on Fig. 16.10, which of the following statements is wrong? (A) The elastic properties of elastin can be described by Hooke's law. (B) Collagen shows a non–linear stress–strain behaviour. (C) Up to strains of about 60%, elastin

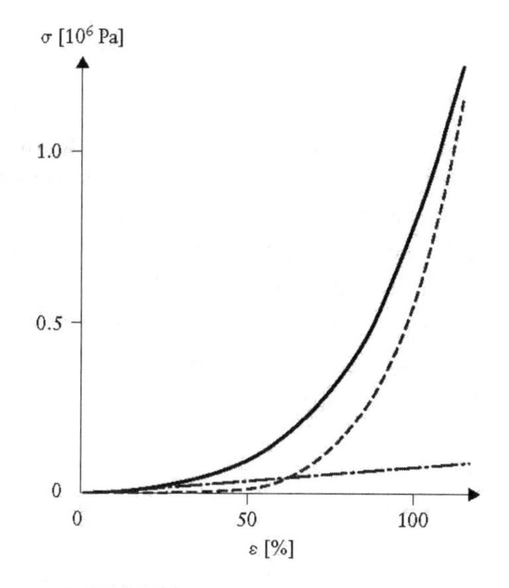

FIGURE 16.10

will dominate the elastic response of the system. (D) The actual blood vessel wall shows non–linear stress–strain behaviour due to its collagen component. (E) All four statements are correct.

Answer: (E)

Question 16.5
What do we imply when we say that a force acting on an object is an elastic force? (A) We imply that we hold the object in a mechanical non–equilibrium position. (B) We imply that the object moves fast in the vicinity of the equilibrium position. (C) We imply that the elastic energy exceeds 10 J. (D) We imply that the total energy of the object is conserved. (E) We imply that the strain is proportional to the stress for the system we study.

Answer: (E)

Question 16.7
We define the positive x–direction as the direction in which an object stretches a spring ($k = 100$ N/m) during a simple harmonic motion. At the instant the object moves through its equilibrium position after having been released at the amplitude point, $x = +A = +0.1$ m, what is the elastic force acting on the object? (A) $F_{elast} = -100$ N. (B) $F_{elast} = -10$ N. (C) $F_{elast} = 0$ N. (D) $F_{elast} = +10$ N. (E) $F_{elast} = +100$ N.

Answer: (C)

Question 16.9
An object is attached to a horizontal spring. It is initially displaced by a distance Δx from the equilibrium position and then released from rest. The object passes the equilibrium position with speed v. From this speed we can calculate the initial displacement Δx by using this formula:

$$(A)\ \Delta x = \sqrt{k}\cdot v \ ;\ (B)\ \Delta x = \sqrt{m}\cdot v$$

$$(C)\ \Delta x = \sqrt{\frac{m}{k}}\,v \ ;\ (D)\ \Delta x = \sqrt{\frac{k}{m}}\,v \qquad \textbf{(1)}$$

$$(E)\ \Delta x = \sqrt{m\cdot k\cdot v}$$

Answer: (C)

Question 16.11
An object is attached to a horizontal spring, which is oriented along the x–axis. The object is initially located at the equilibrium position of the spring and is at rest. Then we hit the object such that it moves along the x–axis, stretching the spring. At what point of its motion along the x–axis does the object reach its lowest kinetic energy? (A) at the equilibrium position, (B) at the positive amplitude position, (C) at midway between the amplitude and equilibrium positions, (D) at the initial position, (E) the kinetic energy is the same at all positions.

Answer: (B)

Question 16.13
We study an object that is attached to a spring and performs a harmonic oscillation on a frictionless horizontal surface. Which parameter set *does not* allow us to calculate the angular frequency of the motion? (A) mass of object and spring constant, (B) frequency of the motion, (C) period of the motion, (D) mass of the object and amplitude of the motion.

Answer: (D)

Question 16.15
An object is attached to a horizontal spring and moves along the x–axis. It is initially displaced to the positive amplitude point, $x = +A$. At that point its elastic potential energy is E_1. Next the object is moved to the opposite amplitude point at $x = -A$. Now its elastic potential energy is E_2. The following relation holds for E_1 and E_2: (A) $E_2 = E_1$. (B) $E_2 = -E_1$. (C) $E_2 = 2\cdot E_1$. (D) $E_2 = -2\cdot E_1$. (E) None of the above.

Answer: (A)

Question 16.17
We replace an object on a spring with one that has four times the mass. How does the frequency of the system change? (A) by a factor 0.25, (B) by a factor 0.5, (C) it remains unchanged, (D) by a factor of 2, (E) by a factor of 4.

Answer: (B). Eq. [16.50] applies:

$$f = \frac{\omega}{2\cdot\pi} = \frac{1}{2\cdot\pi}\sqrt{\frac{k}{m}} \qquad \textbf{(2)}$$

Question 16.19
Does the acceleration of a simple harmonic oscillator remain constant during its motion? Is it ever zero?

Answer: No, the acceleration changes with the restoring force $F_{elast} = -k \cdot (x - x_{eq}) = m \cdot a$. The acceleration is zero at the equilibrium position, where the restoring force is also zero.

Question 16.21
Determine whether the following vectors can point in the same direction during a simple harmonic motion:
(a) displacement and velocity,
(b) velocity and acceleration, and
(c) displacement and acceleration.

Answer to part (a): yes; e.g., during the quarter cycle after the object passes through the equilibrium position.

Answer to part (b): yes; e.g., during the quarter cycle after passing through the two amplitude points.

Answer to part (c): no.

ANALYTICAL PROBLEMS

Problem 16.1. For the graph in Fig. 16.1, express the force (in % of the maximum force) as a mathematical function of the sarcomere length l (in μm) for the linear segments in the interval
(a) 2.2 μm $\leq l \leq$ 3.2 μm,
(b) 2.0 μm $\leq l \leq$ 2.2 μm, and
(c) 1.4 μm $\leq l \leq$ 1.65 μm.

Solution to part (a): The general mathematical formula to describe a linear dependence of y on x is:

$$y = a \cdot x + b \quad \textbf{(3)}$$

in which a (the slope) and b (the intercept of the y–axis) are constant. Two pairs of values for l and $F(l)$ are needed to identify the two constants. The two values for l should be chosen as far apart as possible to minimize the uncertainty in the values calculated for the constants.

From the graph in Fig. 16.1 we choose for the segment in the interval 2.2 μm $\leq l \leq$ 3.2 μm the data pairs l_1 = 2.2 μm with $F(l_1)$ = 100% and l_2 = 3.2 μm with $F(l_2)$ = 32%. Writing the linear formula in the form

$$F = a \cdot l + b \quad \textbf{(4)}$$

with F in percent and l in μm, we find:

(I)	$100 = a \cdot 2.2 + b$
(II)	$32 = a \cdot 3.2 + b$
(I) – (II)	$68 = -1.0 \cdot a$

Thus, a = – 68 %/μm. Substituting this result in either (I) or (II) we obtain b = 250%.

Solution to part (b): Since the line segment in the length interval of 2.0 μm $\leq l \leq$ 2.2 μm is horizontal, we find without calculation that a = 0 %/μm and b = 100%.

Solution to part (c): This part is solved in analogy to part (a). We find a = + 265 %/μm and b = – 344.5%.

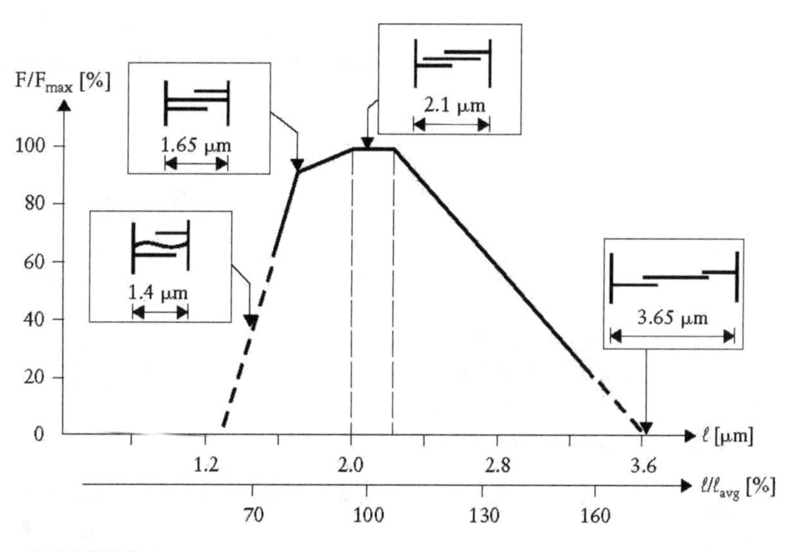

FIGURE 16.1

Problem 16.3
We determine an upper limit of the maximum height of building construction on Earth. This limit is due to the maximum stress in the building material prior to rupture. For steel of density $\rho = 7.9$ g/cm^3 the maximum stress is $\sigma = 2.0 \times 10^8$ Pa. Note that the pressure in the steel at the ground level may not exceed the maximum stress.

Solution: The note at the end asks us to compare the pressure at ground level with the maximum stress with the condition $p \le \sigma$. The pressure is given by the force per unit area, with the force acting at the ground level of the building given by its weight. The weight in turn is given as the mass (obtained from the density and the volume), multiplied with g. Using h for the height of the building with $h = V/A$, in which V is the volume of the building and A its horizontal cross–sectional area, we find:

$$p = \frac{W}{A} = \frac{\rho \cdot V \cdot g}{A} = \rho \cdot h \cdot g \le \sigma \qquad (5)$$

which yields for h with the given values:

$$h \le \frac{\sigma}{\rho \cdot g} = \frac{2 \times 10^8 \text{ Pa}}{\left(7.9 \times 10^3 \frac{\text{kg}}{\text{m}^3}\right)\left(9.8 \frac{\text{m}}{\text{s}^2}\right)} = 2580 \text{ m} \qquad (6)$$

The tallest free–standing structure is the CN Tower in Toronto, Canada, which was built in 1975 and is 554 m high. Since 1996, the tallest office building stands in Kuala Lumpur, Malaysia, and is 452 m high.

Problem 16.5
The four tires of a car are inflated to a gauge pressure of 2.0×10^5 Pa. Each tire is in contact with the ground with an area $A = 240$ cm^2. Determine the weight of the car.

Solution: $W = 1.9 \times 10^4$ N

Problem 16.7
Assume that you have no metre stick but a precise clock. You can then measure the height of buildings by attaching an object to the end of a massless string, with the string pivoted at the top of the structure and the object at the bottom.
(a) If the object swings 10 times back and forth in 110 s, what is the height of the structure?
(b) If you change the object on the massless string to one with double the mass, how does the answer given in part (a) change?

Solution to part (a): We use a simple pendulum for which its length equals the height of the structure. We isolate this length in the formula for the period of a simple pendulum:

$$L = \frac{g \cdot T^2}{4 \cdot \pi^2} = \frac{\left(9.8 \frac{\text{m}}{\text{s}^2}\right)\left(\frac{110 \text{ s}}{10}\right)^2}{4 \cdot \pi^2} = 30 \text{ m} \qquad (7)$$

We used $T = 110$ s/10 because swinging once back and forth corresponds to a full period.

Solution to part (b): Since the mass of the object attached to the string does not enter Eq. [7], varying the mass has no effect on the outcome of the experiment.

Problem 16.9
(a) What is the total energy of the system in Problem 16.8?
(b) What is the elastic potential energy of this system when the object is halfway between the equilibrium position and its turning point?

Solution to part (a): We obtain the total energy from the elastic energy at the amplitude point:

$$E_{total} = \frac{1}{2} k \cdot A^2 = \frac{1}{2}\left(80 \frac{\text{N}}{\text{m}}\right)(0.1 \text{ m})^2 = 0.4 \text{ J} \qquad (8)$$

Solution to part (b): The elastic potential energy at any given displacement from the equilibrium position Δx follows from Eq. [16.34]:

$$E_{elast} = \frac{1}{2} k \left(x - x_{eq}\right)^2 \qquad (9)$$

Substituting the position at the halfway point, $\Delta x = A/2$, we find:

$$E_{elast} = \frac{1}{2} k \left(\frac{A}{2}\right)^2 = \frac{k}{8} A^2 = 0.1 \text{ J} \qquad (10)$$

The kinetic energy at the same point is the difference between the total energy and the elastic potential energy:

$$E_{kin} = E_{total} - E_{elast} = 0.3 \text{ J} \qquad \textbf{(11)}$$

Problem 16.11

An object has a mass of 250 g. It undergoes a simple harmonic motion. The amplitude of that motion is 10 cm and the period is 0.5 s.

(a) What is the spring constant (assuming that the spring obeys Hooke's law)?

(b) What is the maximum magnitude of the force that acts on the object?

<u>Solution to part (a)</u>: We use the relations between angular frequency, frequency and period (Eqs. [16.48] and [16.49]) to relate the spring constant to the period of the vibrational motion:

$$T = \frac{1}{f} = \frac{2 \cdot \pi}{\omega} = 2 \cdot \pi \sqrt{\frac{m}{k}} \qquad \textbf{(12)}$$

Isolating the parameter k we obtain:

$$k = m\left(\frac{2 \cdot \pi}{T}\right)^2 = 0.25\ kg\left(\frac{2 \cdot \pi}{0.5 \text{ s}}\right)^2 = 40\ \frac{\text{N}}{\text{m}} \qquad \textbf{(13)}$$

<u>Solution to part (b)</u>: The maximum force acts on the object when it is at the largest displacement from the equilibrium position, i.e., when it is at $\Delta x = A$. At that point we find with the spring constant obtained in part (a):

$$F_{\max} = k \cdot A = \left(40\ \frac{\text{N}}{\text{m}}\right) 0.1 \text{ m} = 4.0 \text{ N} \qquad \textbf{(14)}$$

Problem 16.13

An object of mass m starts from rest and slides a distance $d = 10$ cm, down a frictionless inclined plane of angle $\theta = 50^0$ with the horizontal, as shown in Fig. 16.32. It then attaches to the end of a relaxed spring with spring constant $k = 100$ N/m. By what length does the object compress the spring by the time it comes (momentarily) to rest?

<u>Solution:</u> We can express the result only in formula form because the mass m is not given:

$$m\, g\, d \sin\theta = \frac{1}{2} k A^2 - m\, g\, A \sin\theta \qquad \textbf{(15)}$$

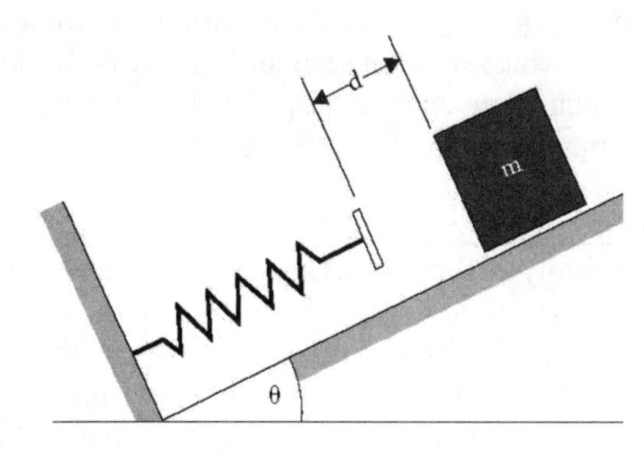

FIGURE 16.32

Problem 16.15

In Fig. 16.32, the object of mass $m = 2.0$ kg is brought into contact with the spring. The object and spring are at rest with the spring compressed by 10 cm. If $\theta = 30^0$, what is the spring constant?

<u>Solution:</u> Since the object–spring system is in equilibrium after the spring has been compressed, then we know from Newton's Second Law that the sum of the forces acting on the object equals zero. Following Example 3.7 in the text, we choose x– and y–axes such that the positive x–direction is uphill and the positive y–direction is upward perpendicular from the inclined surface. The spring force then acts in the positive x–direction, while a component of the object's weight acts in the negative x–direction. These two components must be equal:

$$k\left(x - x_{eq}\right) = m \cdot g \cdot \sin(\theta) \qquad \textbf{(16)}$$

where $\theta = 30^0$ is the angle of the incline. We can find the spring constant as $k = 98$ N/m.

Problem 16.17

An object of mass $m = 0.3$ kg is attached to a horizontal spring. Its position varies with time as:

$$x = (0.25 \text{ m}) \cos(0.4 \cdot \pi \cdot t) \qquad \textbf{(17)}$$

Find

(a) the amplitude of its motion,

(b) the spring constant,

(c) the position at $t = 0.3$ s, and

(d) the speed of the object at $t = 0.3$ s.

<u>Solution to part (a)</u>: The position of an object on a spring with time has the functional form given in Eq. [16.47]. Comparing that equation with the given equation yields $A = 0.25$ m.

Solution to part (b): Again comparing the same two equations, which we can use along with the definition of the angular frequency in Eq. [16.48], we find k = 0.47 N/m.

Solution to part (c): Δx = 0.23 m. Note here that you must multiply all three terms inside the brackets (a factor 0.4, π, and the time 0.3 s) before you take the cosine of that number. Also, you must have your calculator in "radians" mode because the argument of the cosine function is in radians. If your answer was 0.25 m, that is likely because your calculator is in "degrees" mode.

Solution to part (d): We can use conservation of energy to find the speed of the object when we know its position. First, we know that the total energy of the system is given by Eq. [16.39]. At t = 0.3 s we know Δx from part (c). At that instant the total energy has a kinetic and an elastic potential contribution with the speed the only variable. We solve to find v = 0.14 m/s.

Problem 16.19

Fig. 16.34(a) shows a fly and Fig. 16.34(b) shows a simplified model for the insect's moving wings during flight. The wing is pivoted about the outer chitin capsule. The end of the wing lies l_1 = 0.5 mm inside the insect's body and moves up and down by 0.3 mm. We use an effective spring constant of k = 0.74 N/m for the elastic tissue in the insect's body surrounding the end of the wing, and an effective mass of 0.3 mg for the wing. The wing motion is described as a vibration of the end of the wing attached to a spring (elastic tissue).

(a) With what frequency do the wings of the insect flap during flight?

(b) What is the maximum speed of the inner end of the wing?

(c) What is the maximum speed of the outer tip of the wing if the wing is treated as a rigid object?

Use l_2 = 1.4 cm.

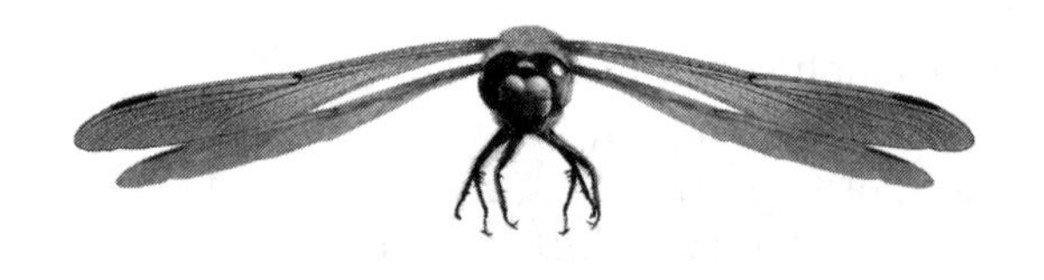

Figure 16.34(a)

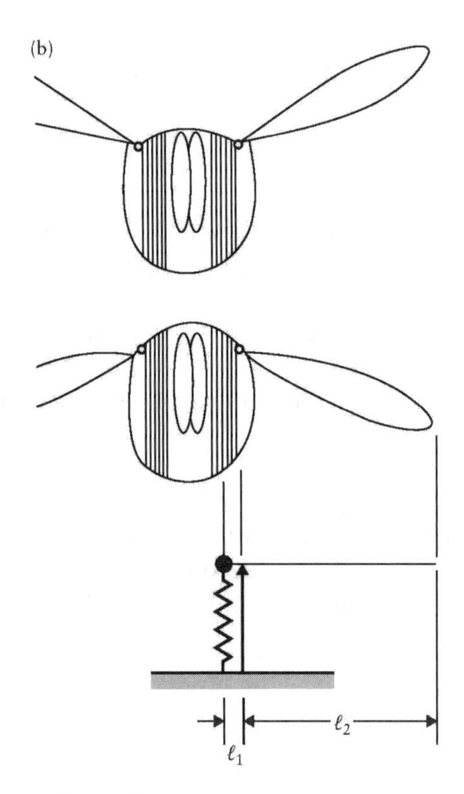

FIGURE 16.34

Solution to part (a): The motion of the wing is described as a simple harmonic motion. We use Eq. [2] to calculate the frequency:

$$f = \frac{1}{2 \cdot \pi}\sqrt{\frac{k}{m}} = \frac{1}{2 \cdot \pi}\sqrt{\frac{0.74 \text{ N}/m}{0.3 \times 10^{-6} \text{ kg}}} = 250 \text{ Hz} \quad \textbf{(18)}$$

This value is a good approximation for many insects, e.g. honey bees flap their wings with a frequency of about 250 Hz and mosquitos with a frequency higher than 500 Hz. The range of frequencies observed in nature is rather wide, with 4 Hz for butterflies and up to 1000 Hz for some gnats.

Solution to part (b): Using Eq. [16.44] we find for the maximum speed of the tip of the inner end of the wing, which has a vertical amplitude of 0.3 mm:

$$v_{max} = \sqrt{\frac{k}{m}}A = \sqrt{\frac{0.74 \text{ N}/m}{0.3 \times 10^{-6} \text{ kg}}}\, 0.3 \times 10^{-3} \text{ m} = 0.5 \frac{\text{m}}{\text{s}} \quad \textbf{(19)}$$

Solution to part (c): We treat the wing as a rigid body rotating about the pivot point. Fig. 16.36 allows us to relate geometrically the maximum speed of the inner end of the wing with the maximum speed at the outer tip:

$$\frac{v_{max,\,in}}{v_{max,\,out}} = \frac{l_1}{l_2} = \frac{0.5\text{ mm}}{14\text{ mm}} = 0.035 \quad \textbf{(20)}$$

which yields:

$$v_{max,\,out} = 28 \cdot v_{max,\,in} = 14\ \frac{\text{m}}{\text{s}} \quad \textbf{(21)}$$

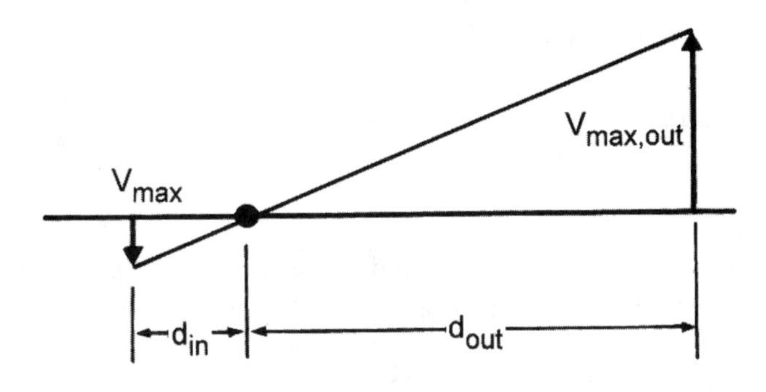

Figure 16.36

Problem 16.21

A pogo stick is a toy that stores energy in a spring with typical spring constant k = 25000 N/m. Fig. 16.35(b) shows a child at three different instances when playing with a pogo stick. At position $d_1 = -10$ cm (panel (A)), the spring compression is at a maximum and the child is momentarily at rest. At position $d = 0$ (panel (B)), the spring is relaxed and the child is moving upward. At position d_2 (panel (C)), the child reaches the highest point of the jump and is momentarily at rest. Assume that the combined mass of the child and the pogo stick is 25 kg.
(a) Calculate the total energy of the system if we choose the gravitational potential energy to be zero at $d = 0$.
(b) Calculate d_2.
(c) Calculate the speed of the child at $d = 0$.
(d) Calculate the acceleration of the child at d_1.
(e) *For those interested*: should any of these results be a matter of concern to the parents?

Solution to part (a): If we choose the gravitational potential energy to be zero at $d = 0$, then the total energy of the system will be the sum of the spring elastic and gravitational potential energy, E_{total} = 101 J. Note that the kinetic energy of the system is zero because it is momentarily at rest when at the pogo stick's maximum

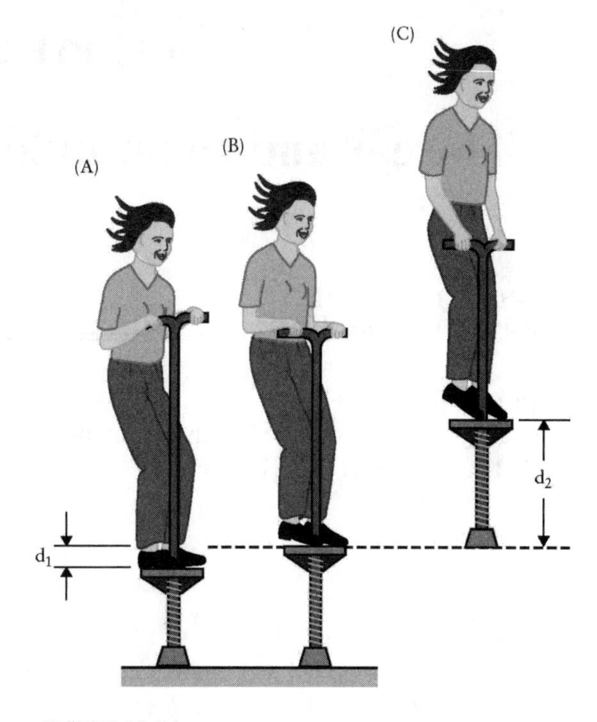

FIGURE 16.35

compression.

Solution to part (b): At the highest point of the child's jump the spring is completely uncompressed and no energy is stored in elastic potential energy. Also, the system is again momentarily at rest so it has no kinetic energy. This means the total energy, which is the same value as in part (a) due to conservation of energy, is equal to the gravitation potential energy. This leads to the result of d_2 = 41 cm.

Solution to part (c): At $d = 0$, the speed of the child can again be found through conservation of energy. At this equilibrium position, the elastic potential energy is zero and the gravitational potential energy is also zero by our original definition in part (a). Therefore all the energy is kinetic energy, which yields v = 2.9 m/s.

Solution to part (d): The acceleration of the child at d_1 can be found by examining the forces acting on him. There will be a gravitational force acting downward which is equal to the child's weight and a spring force acting upward which follows Hooke's law. Together these forces sum to give the upward acceleration experienced by the child. We use Newton's second law to find this acceleration as a = 90.2 m/s².

Solution to part (e): The child is momentarily exposed to more than nine times the gravitational acceleration!

CHAPTER SEVENTEEN

The ear and communication: Longitudinal waves

MULTIPLE CHOICE AND CONCEPTUAL QUESTIONS

Question 17.1
The frequency of a sound wave has the following unit: (A) s, (B) 1/s, (C) m/s, (D) s^2, (E) $1/s^2$.

Answer: (B)

Question 17.3
We compare again two sound waves in air at room temperature. Wave II has twice the frequency of wave I. The following relation holds between their wavelengths: (A) $\lambda_I = \lambda_{II}$, (B) $\lambda_I > \lambda_{II}$, (C) $\lambda_I < \lambda_{II}$, (D) Such a conclusion cannot be drawn with the given information.

Answer: (B)

Question 17.5
We frequently modelled tendons as massless strings. Why is this not a useful model when describing waves on strings?

Answer: A wave on a massless string would have an infinite speed because the mass per length unit of the string is zero.

Question 17.7
A sound source I generates sound with twice the frequency of sound source II. Compared to the speed of sound of source I the speed of sound of source II is (A) twice as fast, (B) half as fast, (C) four times as fast, (D) one–fourth as fast, (E) the same.

Answer: (E)

Question 17.9
The intensity level of a sound is reported in unit decibel (dB). How does IL change if we increase a sound intensity by a factor 10? (A) It remains unchanged, (B) it increases by 1 dB to 2 dB, (C) it increases by 2 dB to 20 dB, (D) it increases by 20 dB to 200 dB, (E) it decreases.

Answer: (C); $\Delta IL = 10$ dB

Question 17.11
When sound is absorbed in a medium, its intensity level IL decreases with distance travelled through the medium x as: (A) IL $\propto e^{-\beta \cdot x}$, (B) IL $\propto -x$, (C) IL $\propto \beta$, (D) IL $\propto \ln(-x)$, (E) none of the above. *Note*: β is a constant.

Answer: (B)

Question 17.13
As a wave and its reflected wave move through each other in a tube that is aligned with the x–axis, there is an instant when the gas in the tube shows no displacement from equilibrium, $\xi = 0$ for all positions x. At that instant, where is the energy carried by the wave?

Answer: In the kinetic energy of the gas elements.

Question 17.15
A tube is initially filled with air (c_{air} = 340 m/s), then with water (c_{water} = 1500 m/s). How does the frequency of the first harmonic change for the tube? (A) no change, (B) it increases, (C) it decreases.

Answer: (B)

Question 17.17
Ultrasound cannot be heard by humans because: (A) its intensity is too low, (B) its frequency is too low, (C) its amplitude is too high, (D) its pressure variations are too high, (E) its frequency is too high.

Answer: (E)

Question 17.19
A moth flies along a path perpendicular to the flight path of a bat. While the moth is within a narrow range of angles in front of the bat, the bat detects a reflected frequency that is (A) less than its emitted frequency, (B) the same as its emitted frequency, (C) more than its emitted frequency, (D) no longer in the range it can hear, (E) in

a range that attracts dogs like a dog whistle.

Answer: (B)

Question 17.21
You are moving toward a stationary wall while emitting a sound. Is there a Doppler shift in the echo you hear? If so, is it the case of a moving source or the case of a moving receiver?

Answer: This corresponds to the case in which source and receiver move, Eqs. [17.87] and [17.88] apply.

ANALYTICAL PROBLEMS

Problem 17.1
A wave with frequency 5.0 Hz and amplitude 40 mm moves in the positive x–direction with speed 6.5 m/s. What are
(a) the wavelength,
(b) the period, and
(c) the angular frequency?
(d) Write a formula for the wave.

Solution to part (a): The wavelength is obtained from the relation of the speed of the wave, the wavelength and the frequency $v_{wave} = \lambda \cdot f$:

$$\lambda = \frac{v_{wave}}{f} = \frac{6.5\ \text{m/s}}{5.0\ \text{Hz}} = 1.3\ \text{m} \quad \textbf{(1)}$$

Solution to part (b): The relation between period and frequency of a vibration is applicable to waves:

$$T = \frac{1}{f} = \frac{1}{5.0\ \text{Hz}} = 0.20\ \text{s} \quad \textbf{(2)}$$

Solution to part (c): The relation between frequency and angular frequency remains unchanged as introduced for vibrations:

$$\omega = 2 \cdot \pi \cdot f = 2 \cdot \pi \cdot 5.0\ \text{Hz} = 31\ \frac{\text{rad}}{\text{s}} \quad \textbf{(3)}$$

Solution to part (d): The harmonic wave function is given in Eq. [17.20]:

$$\xi = A \sin(\omega \cdot t - \kappa \cdot x) \quad \textbf{(4)}$$

The amplitude is given in the problem text. The angular frequency has been calculated in part (c). Thus, we are missing the wave number, which we determine from $\kappa = 2 \cdot \pi/\lambda$ with the wavelength taken from part (a):

$$\kappa = \frac{2 \cdot \pi}{\lambda} = \frac{2 \cdot \pi}{1.3\ \text{m}} = 4.8\ \text{m}^{-1} \quad \textbf{(5)}$$

This allows us to write the wave function:

$$\xi = (0.04\ m) \sin\left(\left\{31.4\ \frac{\text{rad}}{\text{s}}\right\} \cdot t - \left\{4.84\ \frac{1}{\text{m}}\right\} \cdot x\right) \textbf{(6)}$$

Problem 17.3
The range of frequencies heard by the healthy human ear stretches from about 16 Hz to 20 kHz. What are the corresponding wavelengths of sound waves at these frequencies?

Solution: The problem contains an implicit assumption that the hearing process takes place in air of typical environmental conditions. Using $T = 20^0$C, we obtain the speed of sound from Table 17.1 as $c = 343$ m/s. We use this value to analyse the lower frequency of 16 Hz first:

$$\lambda = \frac{c}{f} = \frac{343\ \text{m/s}}{16\ \text{Hz}} = 21.5\ \text{m} \quad \textbf{(7)}$$

The wavelength at the higher frequency is $\lambda = 1.72$ cm.

Table 17.1

Material	Speed of sound (m/s)	Temperature (K)
Gases		
Air	331	273
Air	343	293
Air	386	373
Liquids		
Water	1400	273
Water	1490	298
Seawater (3.5% salt)	1530	298
Solids and soft matter		
Steel	5940	
Granite	6000	
Human body tissue	1540	310
Vulcanized rubber	55	

Problem 17.5

Fig. 17.44 shows a 25-Hz wave travelling in the x–direction. Calculate
(a) its amplitude,
(b) its wavelength,
(c) its period, and
(d) its wave speed.
Use $L_1 = 18$ cm and $L_2 = 10$ cm.

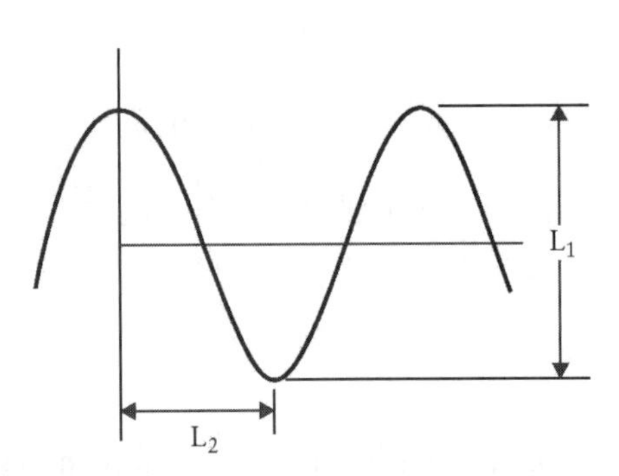

FIGURE 17.44

Solution to part (a): Based on the given diagram, the amplitude is half the distance between the maximum and minimum of the wave, $A = \frac{1}{2} \cdot L_1 = 9$ cm.

Solution to part (b): Based on the given diagram, the wavelength is the distance between successive maxima of the wave, $\lambda = 2 \cdot L_2 = 20$ cm.

Solution to part (c): Since we have the frequency of the wave, we can calculate the period as $T = 0.04$ s.

Solution to part (d): The wave speed is $v = 5.0$ m/s.

Problem 17.7

A piano emits sounds in the range of 28 Hz to 4200 Hz. Find the range of wavelengths at room temperature for this instrument.

Solution: The limits of the range of frequencies 28 Hz < f < 4200 Hz correspond to wavelengths based on Eq. [1]. We find 8.2 cm $\le \lambda \le$ 12.3 m. Note that the lowest frequency has the largest wavelength and vice versa due to the inverse relationship between wavelength and frequency.

Problem 17.9

A person hears an echo 3.0 seconds after emitting a sound. In air of 22°C, how far away is the sound–reflecting wall?

Solution: At a temperature of 22°C, the speed of sound is $c = 344$ m/s. If it takes 3.0 s for a sound wave to travel to a wall and back again (to create an echo) then it must take just 1.5 s for the wave to travel to the wall. Since the wave speed is constant, the distance to the wall must be $d = 516$ m.

Problem 17.11

The only supersonic jet ever used for commercial air travel was the Concorde. It travelled at 1.5 Mach. What was its angle θ, as defined in Fig. 17.45(b), between the direction of propagation of its shock wave and the direction of flight?

Solution: $\sin\theta = 2/3$ or $\theta = 42^0$

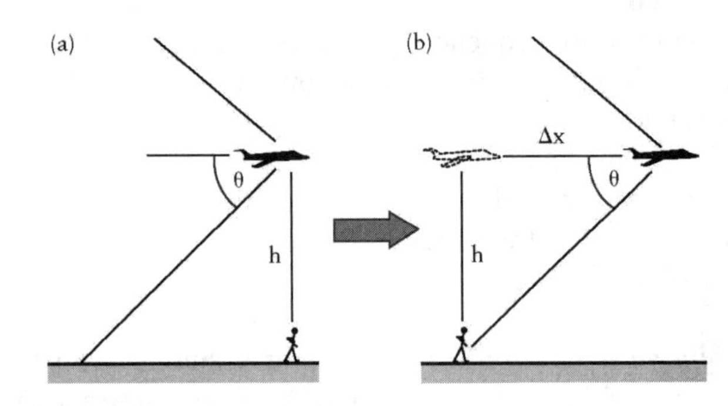

FIGURE 17.45

Problem 17.13

(a) A microphone has an area of 5 cm². It receives during a 4.0-s time period a sound energy of 2.0×10^{-11} J. What is the intensity of the sound?
(b) Using the sound intensity from part (a), what is the variation in pressure in the sound wave, Δp? Use $T = 293$ K and $\rho_{air} = 1.2$ kg/m³.

Solution to part (a): The intensity is defined as the amount of energy transported by a wave per time interval through a plane of unit area which is placed perpendicular to the wave's propagation direction. If we label the intensity I, the energy ΔE, the time interval Δt and the area A, then:

$$I = \frac{1}{A} \cdot \frac{\Delta E}{\Delta t} \quad \textbf{(8)}$$

Using the given values, we find:

$$I = \frac{2 \times 10^{-11}\ \text{J}}{(5 \times 10^{-4}\ \text{m}^2)(4\ \text{s})} = 1 \times 10^{-8}\ \frac{\text{J}}{\text{m}^2 \cdot \text{s}} \qquad \textbf{(9)}$$

Solution to part (b): The pressure variation Δp in the wave is calculated from Eq. [17.22]:

$$\Delta p_{max} = c \cdot \rho \cdot \omega \cdot A \qquad \textbf{(10)}$$

So far, we are not given a value for the amplitude; we find it from re–arranging Eq. [17.33] with the amplitude as the dependent variable, which is then substituted in Eq. [10]:

$$\Delta p = \frac{c \cdot \rho \cdot \omega}{\omega}\sqrt{\frac{2 \cdot I}{c \cdot \rho}} = \sqrt{2 \cdot I \cdot c \cdot \rho} \qquad \textbf{(11)}$$

Eq. [11] allows us now to substitute the given values. The square–root on the right hand side of Eq. [11] reads:

$$\sqrt{2\left(1 \times 10^{-6}\ \frac{\text{J}}{\text{m}^2\ \text{s}}\right)\left(343\ \frac{\text{m}}{\text{s}}\right)\left(1.2\ \frac{\text{kg}}{\text{m}^3}\right)} \qquad \textbf{(12)}$$

which leads to:

$$\Delta p = 2.9 \times 10^{-2}\ \text{Pa} \qquad \textbf{(13)}$$

The speed of sound was taken from Table 17.1, using the given temperature. Note how small these pressure variations in sound waves usually are; the value in Eq. [13] corresponds to 29 mPa!

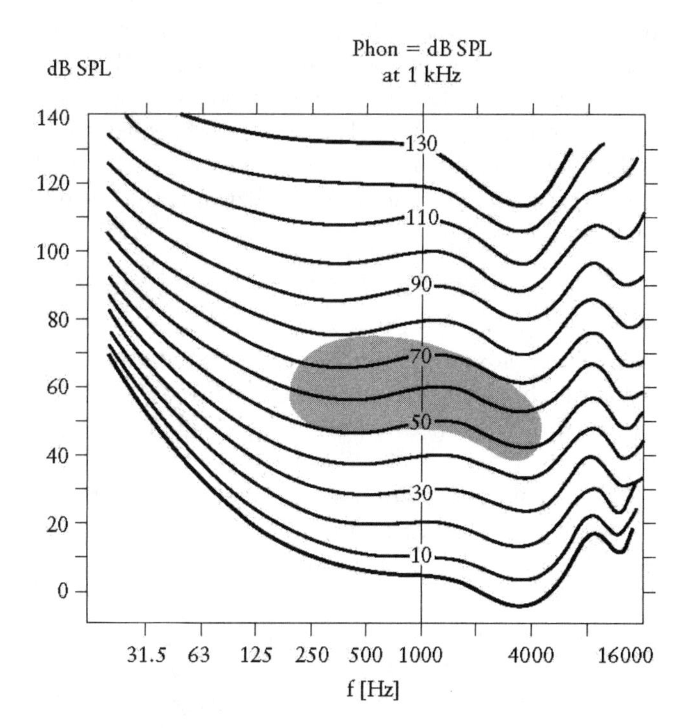

FIGURE 17.33

Problem 17.15

A certain sound has an intensity that is four times the intensity of a reference sound at the same frequency.
(a) What is the difference in the intensity level of the two sounds?
(b) If the reference sound causes a sound perception of 60 phon, what is the sound perception value of the more intense sound?

Solution to part (a): We use the relation between sound intensity and intensity level, IL. Identifying the sound of interest with index 1 and the reference sound with index 2, we write $I_1/I_2 = 4$. This leads to a difference in the intensity levels of the two sounds:

$$IL_1 - IL_2 = 10\left(\log_{10}\frac{I_1}{I_0} - \log_{10}\frac{I_2}{I_0}\right)$$
$$= 10\left(\log_{10}\frac{I_1}{I_2}\right) = 10\log_{10}4 = 6.0\ \text{dB} \qquad \textbf{(14)}$$

Thus, the difference between the two sounds is 6 dB, regardless of the absolute intensity of the sounds.

Solution to part (b): Answering this question is straight forward for a sound of 1 kHz, i.e., at the only frequency where the decibel scale and the phon scale are equal. At that frequency the louder sound has 66 phon. Fig. 17.33 indicates that this relation is approximately correct in the range from 200 Hz to 2000 Hz, i.e., essentially across the frequency range of the human voice during a conversation. At other frequencies, the sound perception would have to be converted to an intensity level using Fig. 17.33, and the louder sound has to be converted back to the phon scale at the same frequency.

Problem 17.17

Two sound waves have intensities of $I_1 = 100\ \text{J/(m}^2 \cdot \text{s)}$ and $I_2 = 200\ \text{J/(m}^2 \cdot \text{s)}$. By how many decibels do the two sounds differ in intensity level?

Solution: The difference in intensity levels may be written as:

$$IL_2 - IL_1 = 10 \log_{10}\left(\frac{I_2}{I_0}\right) - 10 \log_{10}\left(\frac{I_1}{I_0}\right) \quad \textbf{(15)}$$

which we rewrite using the properties of logarithms as:

$$IL_2 - IL_1 = 10 \log_{10}\left(\frac{I_2}{I_0} \cdot \frac{I_1}{I_0}\right) = 10 \log_{10}\left(\frac{I_2}{I_1}\right) \quad \textbf{(16)}$$

Therefore the difference between the intensity levels of the two sounds is 3.01 dB.

Problem 17.19

Ultrasound echolocation is used by bats to enable them to fly and hunt in the dark. The ultrasound used by bats has frequencies in the range from 60 kHz to 100 kHz. We consider a bat that uses an ultrasound frequency of 90 kHz and flies with a speed of 10 m/s. What is the frequency of the echo the bat hears reflected off an insect that moves toward the bat with a speed of 3 m/s?

Solution: Because both the bat and the insect move, we have to combine the two separate Doppler effect cases:

$$f_{receiver} = f_0 \pm \frac{v_{receiver}}{\lambda} = f_0\left(1 \pm \frac{v_{receiver}}{c}\right) \quad \textbf{(17)}$$

which is Eq. [17.82]. Further we use Eq. [17.84]:

$$f_{source} = f_0 \frac{1}{1 \pm \frac{v_{source}}{c}} \quad \textbf{(18)}$$

Different from the application of Doppler ultrasound in medicine (as discussed in the textbook), we have to use different speeds of the source (insect) and the receiver (bat). The combined formula is called the general Doppler effect. With v_{source} the speed of the source and $v_{receiver}$ the speed of the receiver we find:

$$f_{received} = f_{emitted}\left(\frac{1 + v_{receiver}/c}{1 - v_{source}/c}\right) \quad \textbf{(19)}$$

in which $f_{emitted}$ is the frequency emitted by the bat and $f_{received}$ is the frequency received by the bat after reflection off the insect. The + sign in the numerator of Eq. [19] is due to the motion of the receiver *toward* the source and the – sign in the denominator is due to the motion of the source *toward* the receiver. Either of the signs can change if the respective animal moves away from the other; however, the ultrasound emission of bats is focussed in the forward direction and thus, Eq. [19] is meaningless if the bat moves away from the insect. We substitute the given values in Eq. [19]:

$$f_{received} = 9 \times 10^4\ \text{Hz}\left(\frac{1 + \frac{10\ \text{m/s}}{343\ \text{m/s}}}{1 - \frac{3\ \text{m/s}}{343\ \text{m/s}}}\right) \quad \textbf{(20)}$$

$$= 93.4\ \text{kHz}$$

in which we used 343 m/s for the speed of sound in air.

Problem 17.21

A hypothesis says the upper limit in frequency a human ear can hear can be determined by the diameter of the eardrum, which should have approximately the same diameter as the wavelength at the upper limit. If we use this hypothesis, what would be the radius of the eardrum for a person able to hear frequencies up to 18.5 kHz?

Solution: We use the relation of speed of sound, wavelength and frequency. With the given frequency, the diameter of the eardrum can be determined using the speed of sound. Since the speed of sound depends on the temperature of air, we need to make an assumption regarding the temperature of the air in the outer ear. The value lies somewhere between the environmental temperature and the body temperature. The respective speed of sound is then determined from Table 17.1 and Eq. [17.14], with the latter requiring the assumption that air is an ideal gas:

$$c = c_0 \sqrt{1 + \alpha \cdot T} \quad \textbf{(21)}$$

Depending on your pick of temperature, you will use a value of c between 343 m/s (20^{0}C) and 353 m/s (37^{0}C). In the calculation below the speed of sound at 20^{0}C is used.

For the calculation we note that the problem asks for the radius of the eardrum, while the diameter is linked to the wavelength. Thus, we use $2 \cdot r_{eardrum} = \lambda_{max}$:

$$r_{eardrum} = \frac{1}{2} \cdot \frac{c}{f_{max}} = \frac{0.5\left(343\ \frac{m}{s}\right)}{1.85 \times 10^4\ s^{-1}} \quad \textbf{(22)}$$

$$= 9.3 \times 10^{-3}\ m = 9.3\ mm$$

Problem 17.23

If we model the human auditory canal as a tube that is closed at one end and that resonates at a fundamental frequency of 3000 Hz, what is the length of the canal? Use normal body temperature for the air in the canal.

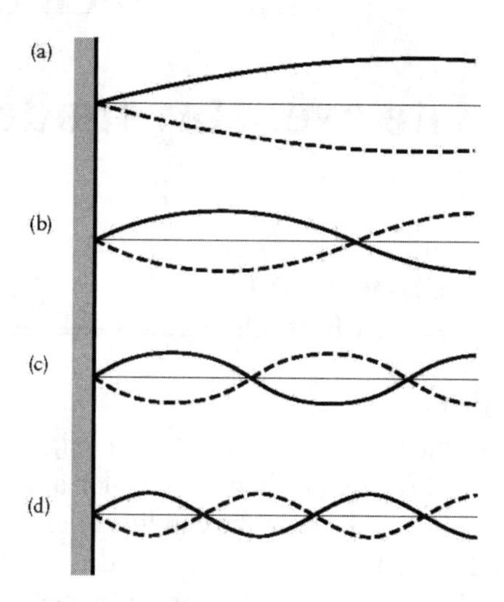

FIGURE 17.16

Solution: From Eq. [17.14] we know that at a temperature of 37^0C, the speed of sound is c = 353 m/s. The wavelength of a 3000 Hz sound wavesis $\lambda = 0.118$ m. The fundamental or first harmonic standing sound wave in a tube that is closed at one end is such that one–quarter of an entire wave exists in the length of the tube, as shown in Fig.17.16. Therefore $L = \lambda/4 = 2.95$ cm.

CHAPTER EIGHTEEN

The eye: Ray model of light (Geometric Optics)

MULTIPLE CHOICE AND CONCEPTUAL QUESTIONS

Question 18.1
A candle produces an image of 0.2 m in height with the wick pointing downward when observed in a flat mirror. What statement is true about the candle in the observer's hand? (A) The candle is 2 cm high and is held upside down. (B) The candle is 2 cm high and is held upright. (C) The candle is 20 cm high and is held upside down. (D) The candle is 20 cm high and is held upright. (E) None of the above.

Answer: (C)

Question 18.3
When you look at yourself in a flat mirror, you see yourself with left and right sides switched, but not upside down. Why? *Hint*: Remember that you are a three–dimensional body. Study the image of the following three vectors: (I) head to foot, (II) left to right hand, and (III) nose to back of head. The remainder of the puzzle is perception of the brain!

Answer: Let's go over the image you see in the mirror in detail. Note that what is at your left side remains at the left side in the image. Further, your head is at the top in the image as it is at the top of the object. Thus, two of the three axis which define a Cartesian coordinate system did not change. The mirror image still does not represent the object in its three–dimensional form as the third direction, the direction toward the mirror, is inverted. What is in front of you toward the mirror is located in the image closer to the mirror than your image.

At this point our brain is unwilling to contemplate the physical facts; it goes beyond comprehension to mirror a person front to back. Instead, the brain interprets the image in a conceptually more acceptable way: because our body has a left–right symmetry the brain concludes that the image is left–right switched.

Question 18.5
Tape a picture of a person on a flat mirror. Approach the mirror to within 20 to 25 cm. Can you focus on the picture and your image at the same time?

Answer: A distance of 20 to 25 cm places the object at the near point of the person (reading distance). The image of the person in the mirror is then apparently 40 to 50 cm from the eye. Both images require different accommodations of the eye that are not possible at the same time.

Question 18.7
A spherical mirror has a focal length of 20 cm. What is the radius of curvature r of the mirror? (A) $r = 0.04$ m, (B) $r = 0.1$ m, (C) $r = 0.4$ m, (D) $r = 1$ m, (E) $r = 4$ m.

Answer: (C)

Question 18.9
An object is placed at $p = +20$ cm in front of a spherical mirror. The image is magnified by a factor of 2 and is inverted. What is the image distance? (A) $q = +0.4$ m, (B) $q = +0.1$ m, (C) $q = -0.1$ m, (D) $q = -0.4$ m, (E) none of the above.

Answer: (A)

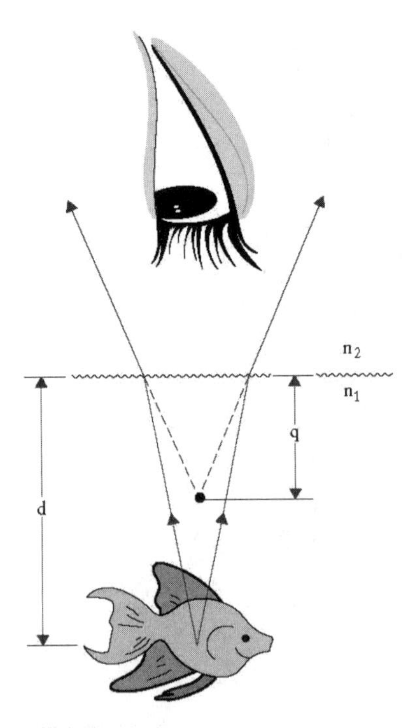

FIGURE 18.21

Question 18.11
Why does the fish in Fig. 18.21 appear closer to the observer than it actually is? (A) Because we look at the fish in the direction perpendicular to the water surface. (B) Because the fish is in water and the observer is in air. (C) Because the index of refraction of the water is smaller than the index of refraction of the air. (D) Because the fish floats toward the surface. (E) Because the water sur-

face is flat, not spherically shaped.

Answer: (B)

Question 18.13

Light is incident at an angle of 45^0 with the vertical on a flat horizontal interface from a vacuum to an unknown type of glass. What is the speed of light in that glass if it travels in the glass with an angle of 27^0 with the vertical? Choose the closest answer. (A) $v = 4 \times 10^8$ m/s, (B) $v = 3 \times 10^8$ m/s, (C) $v = 2 \times 10^6$ km/s, (D) $v = 2 \times 10^5$ km/s, (E) $v = 2 \times 10^4$ km/s.

Answer: (D)

Question 18.15

Sunlight refracts while passing through the atmosphere due to a small difference between the indices of refraction for air and vacuum. We define dawn optically as the instant when the top of the Sun just appears above the horizon, and we define dawn geometrically when a straight line drawn from the observer to the top of the Sun just clears the horizon. Which definition of dawn occurs earlier in the morning?

Answer: The optical definition; due to refraction you see the Sun just below the horizon. The same physics applies at dusk; thus, the optical day is longer than the geometrical day.

Question 18.17

Which of the four bodies in Fig. 18.36, each showing a thin lens made of flint glass ($n = 1.61$), has the smallest refractive power? *Note*: Choose (E) if two bodies tie for the smallest value.

Answer: (A)

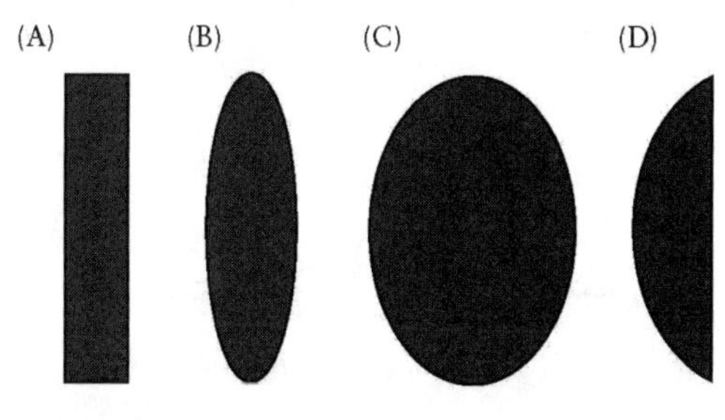

FIGURE 18.36

Question 18.19

A converging lens with focal length $f = 20$ cm is used to view an object 50 cm from the lens. How far from the lens does the object appear? Choose the closest value. (A) 120 cm, (B) 90 cm, (C) 30 cm, (D) 20 cm, (E) 5 cm.

Answer: (C)

Question 18.21

A lens forms an observable magnified image (an image that is larger than the object) if (A) the image is on the same side of the lens as the object; (B) the object is placed at the focal point of the lens; (C) the object is placed closer than twice the focal length in front of the lens; (D) the image distance is smaller than the object distance; (E) none of the above.

Answer: (C)

Question 18.23

Two coaxial converging lenses, with focal lengths f_1 and f_2, are positioned a distance $f_1 + f_2$ apart, as shown in Fig. 18.37. This arrangement is called a beam expander, because it is often used for widening laser beams. If h_1 is the size of the incident beam, the size of the emerging beam is

$$(A)\ h_2 = \frac{f_2}{f_1} h_1 \quad ; \quad (B)\ h_2 = \frac{f_1}{f_2} h_1$$

$$(C)\ h_2 = (f_2 + f_1) h_1 \quad ; \quad (D)\ h_2 = (f_2 - f_1) h_1 \quad \textbf{(1)}$$

$$(E)\ h_2 = (f_2 \cdot f_1) h_1$$

Answer: (A). Note that choices (C) to (E) can immediately be excluded based on a dimensional analysis.

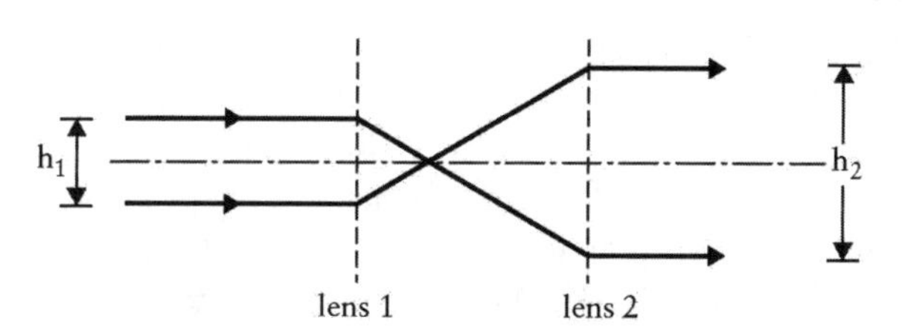

FIGURE 18.37

Question 18.25

A person's face is 30 cm in front of a concave mirror. What is the focal length of the mirror if it creates an upright image that is 1.5 times as large as the actual face? Choose the closest value. (A) 12 cm, (B) 20 cm, (C) 70 cm, (D) 90 cm.

Answer: (D)

Question 18.27

Many nocturnal or crepuscular mammals have eye–shine, like the lesser bush baby in Fig. 18.39: when a bright light is shone into their eyes, a reflection comes back. The reflection is due to a crystalline layer behind the retina, called *tapetum lucidum*. This layer increases the amount of light that passes across the retina, and thereby assists in night vision. What optical system is a good model for the *tapetum lucidum* based on Fig. 18.39? (A) a single, thin lens, (B) a flat mirror, (C) a flat refractive interface, (D) a spherical mirror, (E) a double–lens system like in a microscope.

Answer: (D)

Figure 18.39

Question 18.29

The near point of a particular person is at 50 cm. This is due to a defect of the person's eye; an image forms behind the retina for objects closer than 50 cm, as indi-cated in Fig. 18.31(b). To correct this problem, what refractive power must a corrective lens have (shown in Fig. 18.31(c)) to enable the eye to clearly see an object at 25 cm? (A) $\Re = -2.0$ dpt, (B) $\Re = -1.0$ dpt, (C) $\Re = 0.0$ dpt, (D) $\Re = +1.0$ dpt, (E) $\Re = +2.0$ dpt.

Answer: (E)

ANALYTICAL PROBLEMS

Problem 18.1

When you look at your face in a small bathroom mirror from a distance of 40 cm, the upright image is twice as tall as your face. What is the focal length of the mirror?

Solution: The given distance is the object distance $p = 0.4$ m. The mirror is a spherical mirror since the only other type of mirrors we discuss in the textbook is the flat mirror, and flat mirrors don't generate magnified images. We can use the mirror equation (Eq. [18.8]) to determine the focal length f if we can first identify the image length q. To obtain q we use the information about the magnification of the mirror: $M = +2.0$ since an upright image is associated with a positive magnification (based on the sign conventions in Table 18.1).

Table 18.1

p is positive	Object is in front of mirror (real object)
p is negative	Object is in back of the mirror (virtual object)
q is positive	Image is in front of mirror (real image)
q is negative	Image is in back of the mirror (virtual image)
f and R are positive	Centre of curvature is in front of mirror (concave mirror)
f and R are negative	Centre of curvature is in back of mirror (convex mirror)
M is positive	Image is upright
M is negative	Image is inverted

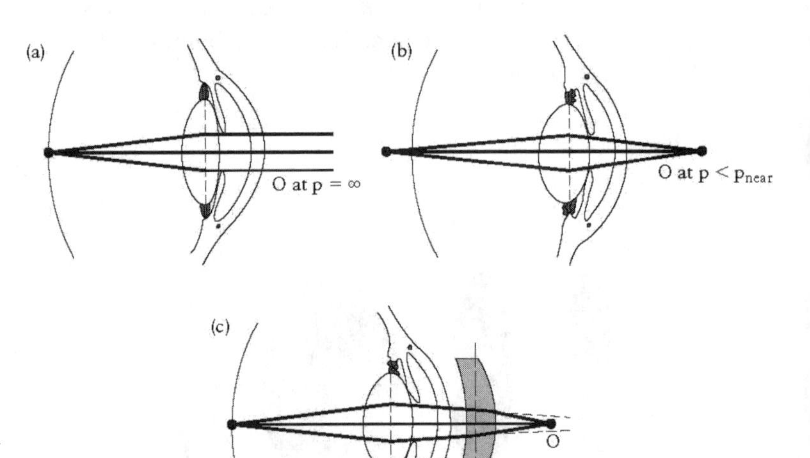

FIGURE 18.31

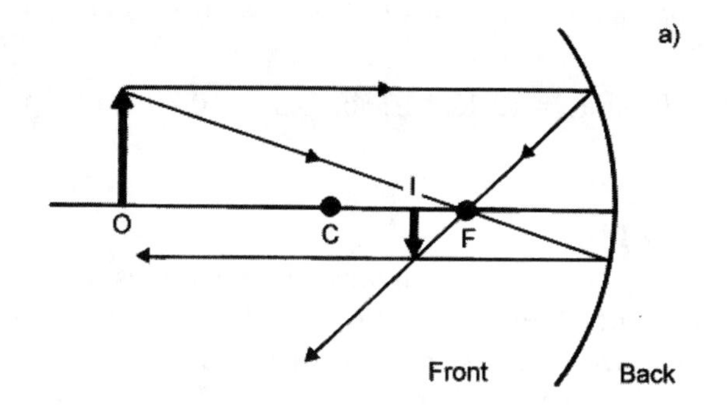

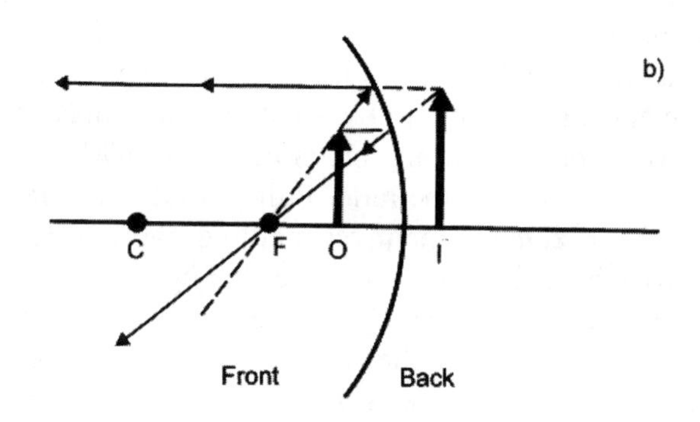

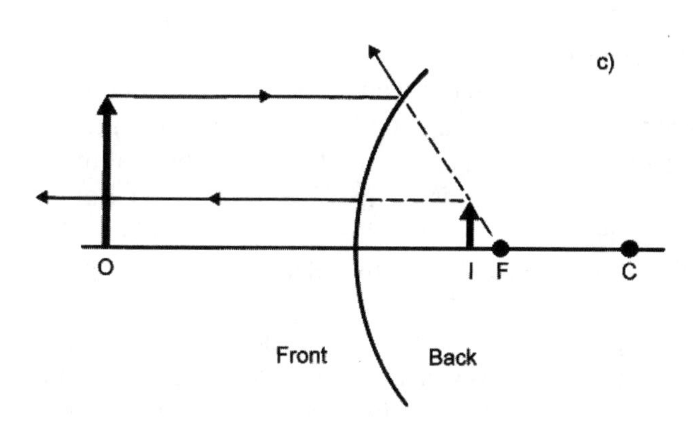

Figure 18.45

Therefore, the equation for the magnification of the mirror (Eq. [18.11]) reads:

$$M = -\frac{q}{p} = 2 \qquad (2)$$

and yields for the image distance:

$$q = -2 \cdot p = -0.8 \text{ m} \qquad (3)$$

With this value for q, the mirror equation reads:

$$\frac{1}{f} = \frac{1}{p} + \frac{1}{q} = \frac{1}{0.4 \text{ m}} - \frac{1}{0.8 \text{ m}} \qquad (4)$$

and yields for the focal length:

$$f = +0.8 \text{ m} \qquad \textbf{(5)}$$

The mirror is concave since f is positive. Note that 80 cm focal length corresponds to a radius of curvature of the mirror of $R = 1.6$ m.

Problem 18.3
Construct the images for the three objects shown in Fig. 18.40.

<u>Solution:</u> The three sketches are shown in Fig. 18.45.

Problem 18.5
A light ray strikes a flat, $L = 2.0$-cm-thick block of glass ($n = 1.5$) in Fig. 18.41 at an angle of $\theta = 30^0$ with the normal.
(a) Find the angles of incidence and refraction at each surface.

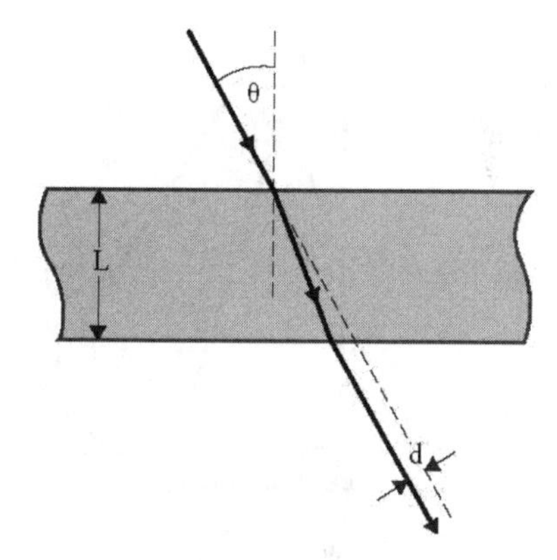

FIGURE 18.41

(b) Calculate the lateral shift of the light ray, d.

<u>Solution to part (a)</u>: We use the law of refraction at the first interface from air to glass, labelling the angles of incidence and refraction θ:

$$\sin\theta_{glass} = \frac{n_{air}}{n_{glass}} \sin\theta_{air} = \frac{1.0}{1.5} \sin 30^0 \qquad (6)$$

which yields:

$$\theta_{glass} = 19.5^0 \qquad (7)$$

For the second interface from glass to air, we have to read the law of refraction in the opposite direction, finding $\theta_{air} = 30^0$. This result is expected as the interface and all the parameters of the light ray exactly mirror the conditions at the first interface.

Solution to part (b): The distance the light ray travels through the glass is obtained geometrically. Fig. 18.46 illustrates the incident beam (which is extended through the glass slab for geometric construction purposes resulting in line AD), the refracted beam (line AC) and the direction across the glass slab perpendicular to the two parallel glass surfaces (line AB, with a given length of 2.0 cm). The distance h we seek is equal to the distance between points A and C. We introduce further the label α for the angle between lines AC and AD, which defines it as $\alpha = \theta_1 - \theta_2$.

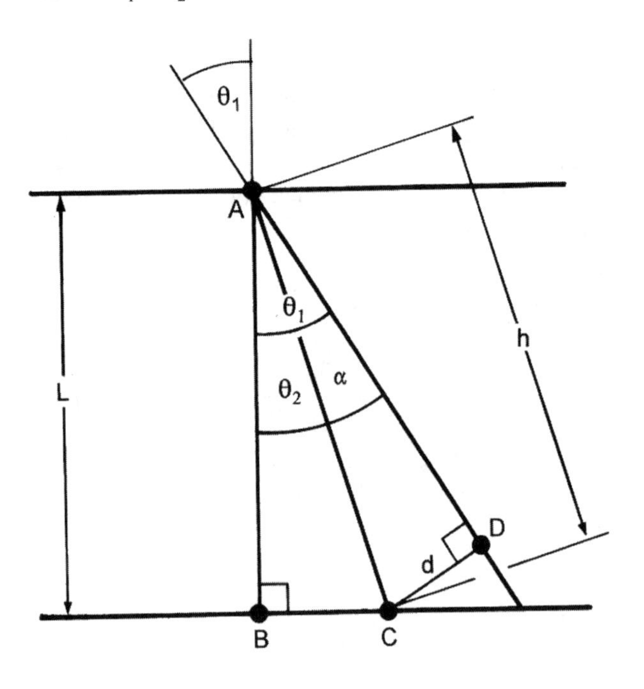

Figure 18.46

Now we use Fig. 18.46 to quantify the length h based on the triangle ABC:

$$\cos\theta_2 = \frac{AB}{AC} = \frac{2\text{ cm}}{h} \quad \textbf{(8)}$$

which yields:

$$h = \frac{2\text{ cm}}{\cos 19.5^0} = 2.12\text{ cm} \quad \textbf{(9)}$$

This does not answer the question, however, because the lateral shift is sought. The lateral shift d is shown in Fig. 18.46 as well. The length d is obtained from the triangle ACD. With $\alpha = \theta_1 - \theta_2 = 30^0 - 19.5^0 = 10.5^0$ we find:

$$d = h \cdot \sin\alpha = (2.12\text{ cm})\sin 10.5^0 = 0.39\text{ cm} \quad \textbf{(10)}$$

Problem 18.7

A light ray travels through air and then strikes the surface of mineral oil at an angle of 23.1^0 with the normal to the surface. What is the angle of refraction if the light ray travels at 2.17×10^8 m/s through the oil?

Solution: The law of refraction is given in the following form, $n_1 \cdot \sin\theta_1 = n_2 \cdot \sin\theta_2$, where medium 1 will be air and medium 2 will be mineral oil. We can rewrite this for the angle in mineral oil by isolating $\sin\theta_2$. Since the index of refraction for any given medium may be calculated as $n = c/v_{light}$, the ratio of the indices of refraction, n_1/n_2, is equal to the inverse ratio of the speeds of light in each medium, v_2/v_1. The speed of light in mineral oil is given and the speed of light in air is approximately equal to the speed of light in vacuum, c. Now we can calculate $\sin\theta_2$, which yields $\theta_2 = 16.5^0$.

Problem 18.9

The laws for refraction and reflection are the same for light and sound. If a sound wave in air approaches a water surface at an angle of 12^0 with the normal of the water surface, what is the angle with the normal of the refracted wave in water? Use for the speed of sound in air 340 m/s and 1510 m/s in water.

Solution: We can use the law of refraction in Eq. [18.17] for sound waves as well as for light waves, re–arranging it with index 1 referring to air and index 2 referring to water. While we don't usually refer to indices of refraction for sound waves, the ratio n_1/n_2 is related to the ratio of the speeds of sound waves in each medium. The relationship is an inverse one, with a medium having slow waves refracting the waves more than a medium having fast waves, exactly the same as for light. We can then make the substitution $n_1/n_2 = v_2/v_1$, which yields the result $\theta_2 = 67.4^0$.

Problem 18.11

A light ray is incident from air onto a glass surface with index of refraction $n = 1.56$. Find the angle of incidence for which the corresponding angle of refraction is one–half the angle of incidence. Both angles are defined with the normal to the surface. For mathematical operations with sine terms see the Math Review *Trigonometry* on p. 566 in the textbook.

Solution: As the light ray passes from air into glass Snell's law (Eq. [18.15]) applies. We are given the fact that the angle of refraction is one–half the angle of incidence, or the angle of incidence is twice the angle of refraction, which we can write as $\alpha = 2 \cdot \beta$. We then refer to the Math Review *Trigonometry* at the end of Chapter 17, which has the theorem:

$$\sin(\alpha + \beta) = \sin\alpha \cdot \cos\beta + \cos\alpha \cdot \sin\beta \quad \textbf{(11)}$$

If we use this theorem as it applies to our problem and let α and β both be equal, then we can write this theorem as $\sin(2 \cdot \alpha) = 2 \cdot \sin\alpha \cdot \cos\alpha$. We can use this formula, often referred to as a double–angle formula, in Snell's law; we find $\beta = 38.7^0$, and $\alpha = 77.4^0$.

Problem 18.13

A converging lens has a focal length $f = 20.0$ cm. Locate the images for the object distances given below. For each case state whether the image is real or virtual and upright or inverted, and find the magnification.
(a) 40 cm;
(b) 20 cm;
(c) 10 cm.

Solution to part (a): The focal length f is positive for a converging lens as stated in Table 18.4.

Table 18.4

p is positive	Object is in front of the lens
p is negative	Object is in back of the lens
q is positive	Image is in back of the lens
q is negative	Image is in front of the lens
R_1 and R_2 are positive	Centre of curvature for each surface is in back of the lens
R_1 and R_2 are negative	Centre of curvature for each surface is in front of the lens
f is positive	Converging lens
f is negative	Diverging lens

To solve the problem, the thin–lens formula from Eq. [4] and the magnification in Eq. [2] are used, with the focal length $f = +\,20.0$ cm. For $p = 40$ cm, we find:

$$\frac{1}{q} = \frac{1}{f} - \frac{1}{p} = \frac{1}{20\ \text{cm}} - \frac{1}{40\ \text{cm}} \quad \textbf{(12)}$$

which yields for the image distance:

$$q = +\,40\ \text{cm} \quad \textbf{(13)}$$

and for the magnification:

$$M = -\frac{q}{p} = -\frac{40\ \text{cm}}{40\ \text{cm}} = -\,1.0 \quad \textbf{(14)}$$

The image is real and inverted. It is located 40 cm behind the lens. In this case, the image has the same size as the object.

Solution to part (b): For $p = 20$ cm, we find:

$$\frac{1}{q} = \frac{1}{f} - \frac{1}{p} = \frac{1}{20\ \text{cm}} - \frac{1}{20\ \text{cm}} \quad \textbf{(15)}$$

i.e., $q = \infty$; therefore, no image is formed. The light rays emerging from the lens travel parallel to each other.

Solution to part (c): For $p = 10$ cm, we find:

$$\frac{1}{q} = \frac{1}{f} - \frac{1}{p} = \frac{1}{20\ \text{cm}} - \frac{1}{10\ \text{cm}} \quad \textbf{(16)}$$

which yields for the image distance:

$$q = -\,20\ \text{cm} \quad \textbf{(17)}$$

and for the magnification:

$$M = -\frac{q}{p} = -\left(\frac{-\,20\ \text{cm}}{10\ \text{cm}}\right) = +\,2.0 \quad \textbf{(18)}$$

The image is upright and virtual. It is located 20 cm in front of the lens. The image is twice as large as the object.

Problem 18.15

Fig. 18.44 shows an object at the left, a lens at the centre (vertical dashed line), and a concave mirror at the right. The respective focal lengths and the distance between lens and mirror are indicated at the bottom of the figure. Construct the image that forms after light from the object has passed through the lens and has reflected off the mirror.

No solution provided. Draw this one on your own, using Fig. 18.44.

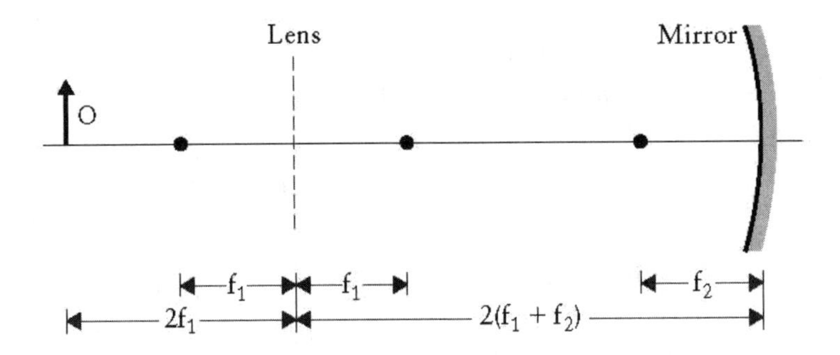

FIGURE 18.44

Problem 18.17
A contact lens is made of plastic with an index of refraction of $n = 1.58$. The lens has a focal length of $f = +25.0$ cm, and its inner surface has a radius of curvature of +18.0 mm. What is the radius of curvature of the outer surface?

Solution: This problem requires the combination of thin–lens formula (Eq. [4]) and lens maker's equation (Eq. [18.56]) in the form:

$$\frac{1}{f} = (n - 1)\left(\frac{1}{R_1} - \frac{1}{R_2}\right) \quad \textbf{(19)}$$

Note that we consider the same medium on both sides of the lens. We isolate the radius R_1 in Eq. [19]:

$$\frac{1}{n - 1}\left(\frac{1}{f} + (n - 1)\frac{1}{R_2}\right) = \frac{1}{R_1} \quad \textbf{(20)}$$

which yields:

$$R_1 = \frac{(n - 1) \cdot f \cdot R_2}{(n - 1) \cdot f + R_2} \quad \textbf{(21)}$$

Substituting the given values in Eq. [21] yields:

$$R_1 = + 1.6 \text{ cm} = + 16 \text{ mm} \quad \textbf{(22)}$$

Problem 18.19
The near point of an eye is 100 cm. A corrective lens is to be used to allow this eye to focus clearly on objects 25 cm in front of it.
(a) What should be the focal length of the lens?
(b) What is the refractive power $\mathfrak{R}$ of the lens?

Solution to part (a): We use the thin–lens formula to determine the focal length of the prescription lens. The text identifies the distance to the object as $p = 25$ cm. The prescription lens must form a virtual image at a distance $q = -100$ cm since the image of the prescription lens serves as the object for the eye's lens. We find therefore:

$$\frac{1}{f} = \frac{1}{p} + \frac{1}{q} = \frac{1}{25 \text{ cm}} + \frac{1}{-100 \text{ cm}} \quad \textbf{(23)}$$

which yields $f = 33.3$ cm.

Solution to part (b): The corresponding refractive power is:

$$\mathfrak{R} = \frac{1}{f} = \frac{1}{0.333 \text{ m}} = + 3.0 \text{ dpt} \quad \textbf{(24)}$$

Problem 18.21
The near point of a patient's eye is 75.0 cm.
(a) What should be the refractive power $\mathfrak{R}$ of a corrective lens prescribed to enable the patient to clearly see an object at 25.0 cm?
(b) When using the new corrective glasses, the patient can see an object clearly at 26.0 cm but not at 25.0 cm. By how many diopters did the lens grinder miss the prescription?

Solution to part (a): The near point at 75.0 cm means that the patient cannot see objects clearly which are closer than that distance from the eye. Reading is therefore a problem, which is likely the reason why the patient came to see an optometrist.

The optometrist prescribes corrective glasses such that the patient can clearly see (e.g. read a text) at a distance of 25 cm. Thus, for an object placed at object distance $p = +25.0$ cm the prescribed lens must form a virtual image at $q = -75.0$ cm at which the patient's eye then can look. We calculate for this situation the focal length of the prescribed lens from the thin–lens formula:

$$\frac{1}{f} = \frac{1}{p} + \frac{1}{q} = \frac{1}{25\ \text{cm}} + \frac{1}{-75\ \text{cm}} \qquad \textbf{(25)}$$

which yields $f = 37.5$ cm and $\Re = 1/f = 2.67$ dpt.

Solution to part (b): We repeat the calculation of part (a) except that we use $p = 26.0$ cm for the object distance for the prescription lens. Substituting this value in Eq. [25] we find for the actual focal length $f = 39.8$ cm and for the actual refractive power $\Re = 2.51$ dpt. Thus, the error is 0.16 dpt.

CHAPTER NINETEEN

The microbial world: Microscopy

MULTIPLE CHOICE AND CONCEPTUAL QUESTIONS

Question 19.1

A person with a near point of 35 cm tries to see a text with small print better by bringing the page closer to the eye. The person will achieve what angular magnification? (A) none; (B) $m = 1$ (no gain); (C) $m < 1$ (the person does worse than the standard man); (D) $m > 1$ but $m < 2$ — a moderate angular magnification is achieved; (E) $m >> 1$ (for this person, this is the way to go to see small objects).

<u>Answer:</u> (C)

Question 19.3

Assume you use a converging lens as a magnifying glass. Initially, you hold the lens far from a page with small print. Then you move the lens closer and closer to the text until the lens lies on the page as shown in Fig. 19.9. What do you observe? (A) The text is always upright, no matter how far the lens is held. (B) The text is initially inverted, then blurs and becomes upright. (C) The text is initially upright, then blurs and becomes inverted. (D) When the magnifying glass is held far enough from the page the text will run from right to left.

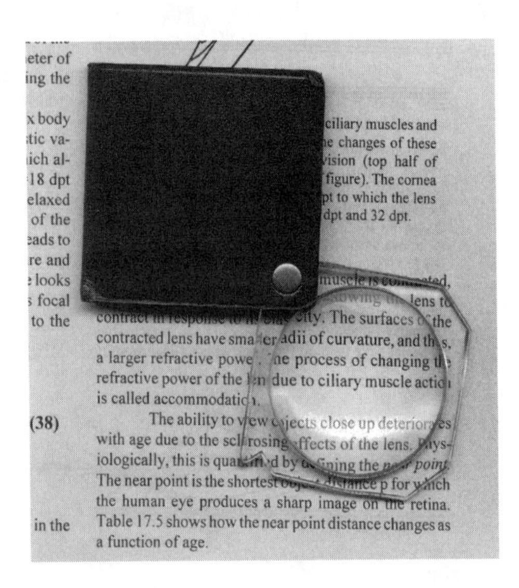

Figure 19.9

<u>Answer:</u> (A). The best way to confirm this result is to either do the experiment if you have a magnifying glass at hand, or to use Fig. 19.9 and logic conjecturing about the purpose of a magnifying glass.

Question 19.5

Which of the following total angular magnifications m_{total} can be selected with a compound microscope with an eyepiece of angular magnification $m_E = 20$ and three switchable objective lenses with magnifications M_O of 10, 30, and 50? (A) $m_{\text{total}} = 100$, (B) $m_{\text{total}} = 300$, (C) $m_{\text{total}} = 500$, (D) $m_{\text{total}} = 700$, (E) none of the four total angular magnifications above can be selected with this microscope.

<u>Answer:</u> (E). You can select $m_{\text{total}} = 200$, 600, and 1000.

Question 19.7

Most commercial microscopes have an additional lens, called the condenser lens, which is located between the light source and the object (see (4) in Fig. 19.1(b)). What

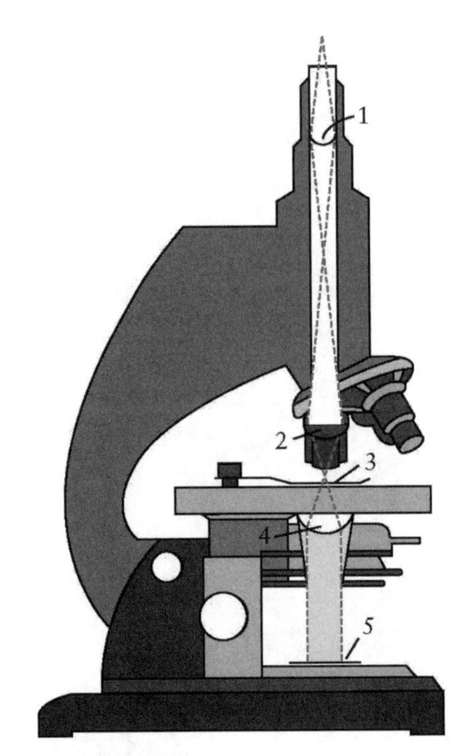

FIGURE 19.1

does this lens do? (A) Enhance the overall angular magnification. (B) Invert the image so that we don't see everything upside down. (C) Substitute for the eyepiece when the eyepiece becomes defective due to poor upkeep of the instrument. (D) Focus light from a light source on the object. (E) Illuminate the image for faster photographic exposure.

Answer: (D)

Question 19.9
You are unhappy with the overall magnification you achieve with your homemade microscope. Which alteration will improve the results? (A) Shorten the distance between the objective lens and the eyepiece. (B) Exchange the objective lens for a lens with a larger focal length. (C) Exchange the eyepiece for a lens with a larger focal length. (D) Loosen up and look through the microscope with a relaxed eye. (E) None of the above.

Answer: (E)

Question 19.11
We study two thin lenses, where lens 1 with $f_1 = 15$ cm is placed a distance of $L = 35$ cm to the left of lens 2 with $f_2 = 10$ cm. An object is then placed 50 cm to the left of lens 1. What is the magnification of the final image taken with respect to the object? *Note*: The magnification you determine is not an angular magnification. (A) $M = 0.6$, (B) $M = 1.0$ (no magnification), (C) $M = 1.2$, (D) $M = 2.4$, (E) $M = 3.6$.

Answer: (C)

ANALYTICAL PROBLEMS

Problem 19.1
A magnifying glass is used to examine the structural details of a human hair. The hair is held 3.5 cm in front of the magnifying glass, and the image is 25.0 cm from the lens.
(a) What is the focal length of the magnifying glass?
(b) What angular magnification is achieved?

Solution to part (a): The text identifies the object and image distances for the magnifying glass as $p = 3.5$ cm and $q = -25.0$ cm. The image distance must be negative since the image is located on the same side of the lens as the object. Using the thin–lens formula we find for the focal length:

$$\frac{1}{f} = \frac{1}{p} + \frac{1}{q} = \frac{1}{3.5\ \text{cm}} + \frac{1}{-25.0\ \text{cm}} \qquad \textbf{(1)}$$

which yields $f = 4.07$ cm.

Solution to part (b): Remember that we have two formulas for the angular magnification, of these Eq. [18.7] applies to the relaxed eye (object at infinite distance). In the current case the formula for the focussed eye is used, which carries an additional term of one:

$$m = 1 + \frac{s_0}{f} = 1 + \frac{25.0\ \text{cm}}{4.07\ \text{cm}} = 7.14 \qquad \textbf{(2)}$$

Problem 19.3
A microscope has an objective lens with $f = 16.22$ mm and an eyepiece with $f = 9.5$ mm. With the length of the microscope's barrel set at 29.0 cm, the diameter of an erythrocyte's image subtends an angle of 1.43 mrad with the eye. If the final image distance is 29.0 cm from the eyepiece, what is the actual diameter of the erythrocyte? *Hint*: Start with the size of the final image. Then use the thin–lens formula for each lens to find their combined magnification. Use this magnification to calculate the object size in the final step.

Solution: This problem shows how calculations for multi–lens systems become sometimes more complicated. You want to recognize that the added complexity is not due to the need of more challenging mathematical skills but is due to the multi–step approach you have to take to find the solution. These steps are associated with treating one lens after the other and determining the location and magnification of intermediate images. Use the next few problems also to develop a strategy of how to tackle problems of this type.

In the current problem we start with the size of the final image which is only given as an angle of observation of $\theta = 1.43$ mrad. We use the geometrical sketch in Fig. 19.12 which illustrates the relation be-

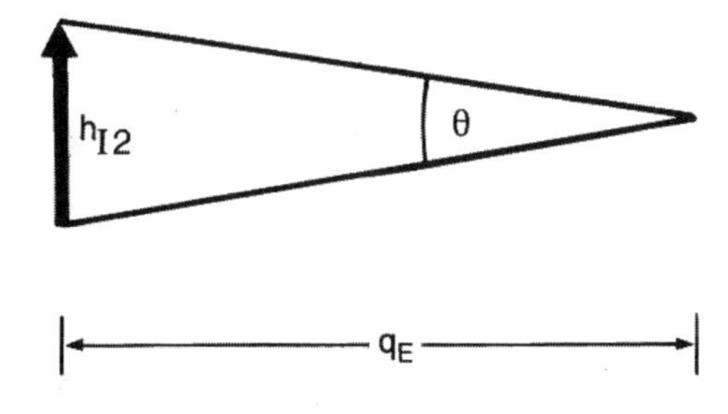

Figure 19.12

tween θ, the height (size) of the final image, h_{I2}, and the distance between image and eyepiece, q_E, given as q_E = 29.0 cm in the text. Noting that θ is very small we write:

$$\sin\theta \cong \theta = 1.43 \times 10^{-3} \text{ rad} = \frac{h_{I2}}{|q_E|} \quad (3)$$

which yields for the image height:

$$h_{I2} = (29 \text{ cm}) \cdot (1.43 \times 10^{-3}) = 4.15 \times 10^{-2} \text{ cm} \quad (4)$$

In the second step the thin–lens formula and the formula for the magnification of a lens are used to find the magnification of each of the two lenses of the microscope. Fig. 19.13 identifies the respective object and image locations and sizes.

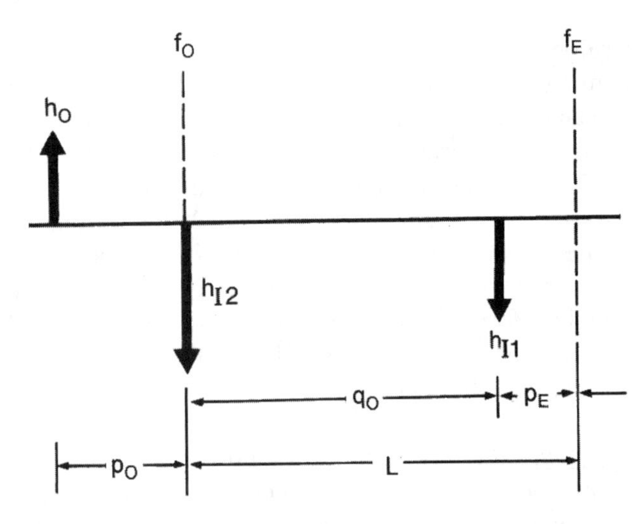

Figure 19.13

• for the eyepiece we determine the object distance from the thin–lens formula:

$$\frac{1}{p_E} = \frac{1}{f_E} - \frac{1}{q_E} = \frac{1}{0.95 \text{ cm}} - \frac{1}{-29 \text{ cm}} \quad (5)$$

which yields p_E = 0.92 cm. This is then substituted in the magnification formula:

$$M_E = -\frac{q_E}{p_E} = -\frac{-29 \text{ cm}}{0.92 \text{ cm}} = 31.5 \quad (6)$$

• for the objective lens we note that $q_O = L - p_E$ as shown in Fig. 19.13. Thus, q_O = (29.0 cm) – (0.92 cm) = 28.08 cm. With this value we use again the thin–lens formula:

$$\frac{1}{p_O} = \frac{1}{f_O} - \frac{1}{q_O} = \frac{1}{1.622 \text{ cm}} - \frac{1}{-28.08 \text{ cm}} \quad (7)$$

which yields p_O = 1.72 cm, which we then substitute in the equation for the magnification:

$$M_O = -\frac{q_O}{p_O} = -\frac{-28.08 \text{ cm}}{1.72 \text{ cm}} = -16.3 \quad (8)$$

Combining the result for both lenses we find the combined magnification of the microscope:

$$M = M_E \cdot M_O = -16.3 \cdot 31.5 = -513.5 \quad (9)$$

Knowing the size of the final image, the total magnification calculated in Eq. [9] allows us to determine the size of the original object:

$$h_O = \frac{h_{I2}}{|M|} = \frac{4.15 \times 10^{-2} \text{ cm}}{513.5} = 0.81 \ \mu\text{m} \quad (10)$$

Problem 19.5
An object is located 20 cm to the left of a converging lens of focal length 25 cm. A diverging lens with focal length 10 cm is located 25 cm to the right of the converging lens. Find the position of the final image.

Solution: Here we have a system of two lenses and we can find the location of the image produced by the first lens and consider that image to be the object for the second lens. For the first lens, f_1 = 25 cm and p_1 = 20 cm. We use the thin–lens equation to find q_1 = – 100 cm. This means that the created image is virtual and located 100 cm to the left of the first lens. Since the two lenses are 25 cm apart, this means the first image is 125 cm to the left of the second lens. We can use this as our new object distance, q_2. Noting that the focal length of this diverging lens will be negative, we then use the thin–lens equation to get q_2 = – 9.3 cm. The position of the final image is 9.3 cm to the left of the second lens, i.e. between the two lenses.

CHAPTER TWENTY

Colour vision: Magnetism and the electromagnetic spectrum

MULTIPLE CHOICE AND CONCEPTUAL QUESTIONS

Question 20.1
In a presentation, you use Fig. 20.7 to describe the magnetic force between two current–carrying wires. Someone in the audience challenges your statement that $\mathbf{F}_{2 \text{ on } 1}$ and $\mathbf{F}_{1 \text{ on } 2}$ are an action–reaction pair of forces. Which of the following statements would you not make in response? (A) The two forces act on different objects. (B) The forces are equal in magnitude. (C) The forces are opposite to each other in direction. (D) The electric currents flow in parallel directions.

Answer: (D)

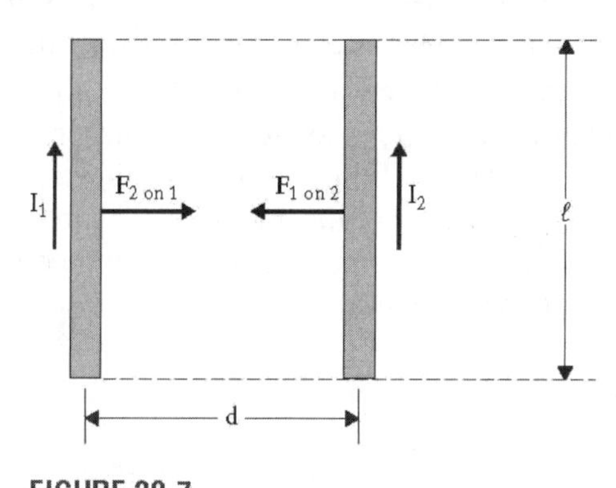

FIGURE 20.7

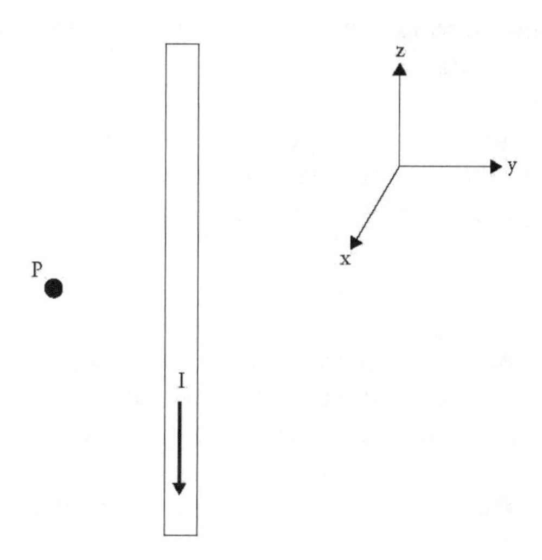

FIGURE 20.38

Question 20.3
Fig. 20.38 shows a long wire that carries a current I. In what direction does the magnetic field point at point P, which lies in a common yz–plane with the wire? Answer the question based on the coordinate system shown in the figure. (A) along the positive x–axis, (B) along the negative x–axis, (C) along the positive y–axis, (D) along the negative y–axis, (E) along the positive or negative z–axis.

Answer: (B)

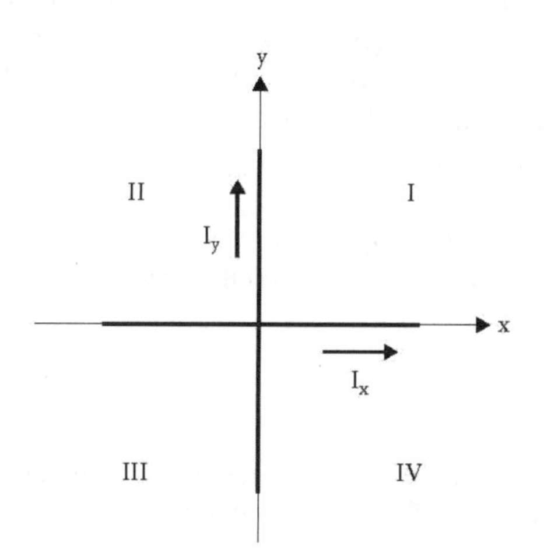

FIGURE 20.11

Question 20.5
Consider Fig. 20.11. In which quadrants does one conductor contribute to the magnetic field in the positive z–direction and the other in the negative z–direction? (A) in quadrants I and II, (B) in quadrants I and III, (C) in quadrants I and IV, (D) in quadrants II and III, (E) in quadrants II and IV, (F) in quadrants III and IV. (G) None of these choices is correct.

Answer: (B)

Question 20.7
In which arrangement is a conductor with a current not subjected to a magnetic force?

Answer: A current flows parallel or anti–parallel to the magnetic field vector.

Question 20.9
We move with the drift speed parallel to the electrons in a current–carrying conductor. Do we measure a zero magnetic field?

Answer: No. The metal's positive core ions appear to move with the drift speed in the opposite direction. Thus, the same magnetic field is observed.

Question 20.11
Parallel conductors exert magnetic forces on each other. What about two current–carrying conductors that are oriented perpendicular to each other?

Answer: There is no net force but a torque acting on the conductors.

Question 20.13
Light travels in the direction that is (A) parallel to the electric field vector, (B) parallel to the magnetic field vector, (C) at an angle of 45^0 with the electric field vector, (D) at an angle of 45^0 with the magnetic field vector, (E) none of the above.

Answer: (E). It propagates in the direction that is perpendicular to both the electric and magnetic field vectors.

Question 20.15
We use a conducting line as a receiving antenna. What should the orientation of this antenna be relative to the antenna that emits electromagnetic waves?

Answer: The two line antennas have to be parallel to each other.

Question 20.17
What is it that actually moves when a light wave travels through outer space?

Answer: Energy, in this case often referred to as a photon (light corpuscle).

Question 20.19
Why does an infrared photograph taken of a person look different from a photograph taken with visible light?

Answer: The intensity profile of light emission from a person in the visible range is different from the intensity profile in the infrared.

Question 20.21
Light is fully polarized if it is reflected off a planar surface and the incoming light ray (A) forms an angle of 60^0 with the normal, (B) forms an angle of 45^0 with the normal, (C) forms an angle of 30^0 with the normal, (D) hits the surface perpendicular, (E) none of the above.

Answer: (E). See Brewster's angle that depends on the index of refraction.

Question 20.23
When white light passes through a prism it is split into the colours of the rainbow. The following feature does not contribute to this observation: (A) White light contains light of all visible frequencies (between 360 nm and 760 nm). (B) The index of refraction of the prism material depends on the wavelength of the light. (C) The prism material displays a non–constant dispersion relation. (D) The speed of light for the various wavelengths varies in the glass body of the prism. (E) The frequencies of the different light components change unequally as the light enters the prism.

Answer: (E)

Question 20.25
Fig. 20.40 shows that our eye is much more sensitive to absolute intensities of green light in comparison to ab-

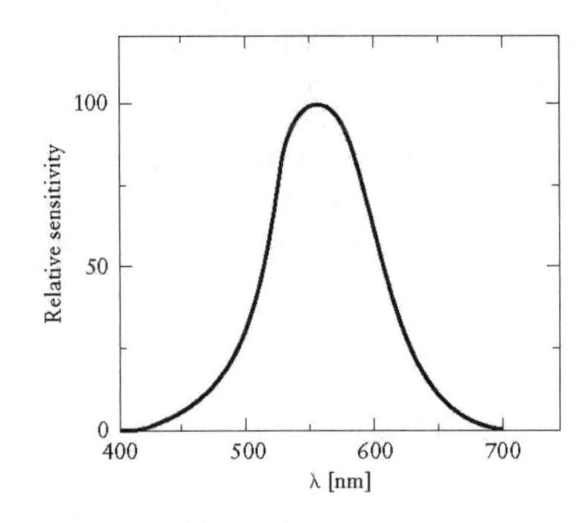

FIGURE 20.40

solute intensities of red light. Why then do the green and red lights of a traffic light still look roughly equally bright?

Answer: If a green street light would emit green light (e.g. in the 500 to 550 nm wavelength range) and a red street light would emit red light (e.g. in the 650 to 700 nm wavelength range) then, indeed, the green light were about twenty times brighter than the red light. However, actual street lights function in a different fashion. The colour of a street light is the result of elimination of certain wavelengths: The light bulb in the street light emits white light and a filter eliminates part of the spectrum to create the final colour impression. As an example, if molecules embedded in the filter absorb in the green part of the spectrum then you see the complementary colour, in this case red. Thus, you see all street light colours about equally intensive as only minor fractions of the white light are removed.

Question 20.27
Why is a rainbow red at the top and blue at the bottom?

Answer: Water and ice particles in the atmosphere refract sunlight toward your eye in the same fashion as a prism. Blue (and violet) light are refracted more strongly than red light.

Question 20.29
A mixed light beam of two colours, X and Y, is sent through a prism. In the prism, the X component is bent more than the Y component. Which component travelled more slowly in the prism?

Answer: The colour travelling more slowly is bent more.

ANALYTICAL PROBLEMS

Problem 20.1
Two long, parallel wires are separated by a distance of $l_2 = 5$ cm, as shown in Fig. 20.42. The wires carry currents $I_1 = 4$ A and $I_2 = 3$ A in opposite directions. Find the direction and magnitude of the net magnetic field
(a) at point P_1 that is a distance $l_1 = 6$ cm to the left of the wire carrying current I_1, and
(b) at point P_2 that is a distance $l_3 = 5$ cm to the right of the wire carrying current I_2.
(c) At what point is the magnitude of the magnetic field zero, i.e., $|\mathbf{B}| = 0$?

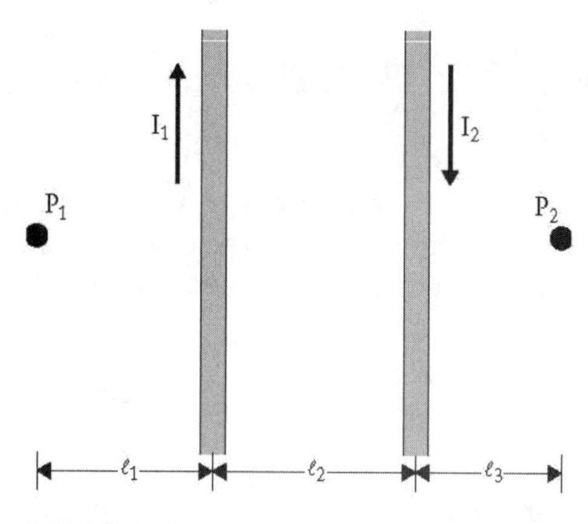

FIGURE 20.42

Solution: We quantify the magnetic field for each of the two wires before answering the three questions. The magnetic field as a function of distance from a current–carrying wire is given in Eq. [20.3]:

$$|\boldsymbol{B}| = \frac{\mu_0}{2 \cdot \pi} \cdot \frac{I}{d} \qquad (1)$$

For the wire carrying I_1 we find:

$$|\boldsymbol{B}_1| = \frac{\left(1.26 \times 10^{-6} \frac{\mathrm{N}}{\mathrm{A}^2}\right)(4\ \mathrm{A})}{2 \cdot \pi \cdot d_1} = \frac{8.0 \times 10^{-7}\ \mathrm{T \cdot m}}{d_1} \qquad (2)$$

in which d_1 is the distance from wire 1. For the wire carrying I_2 we find in the same fashion:

$$|\boldsymbol{B}_2| = \frac{6.0 \times 10^{-7}\ \mathrm{T \cdot m}}{d_2} \qquad (3)$$

in which d_2 is the distance from wire 2.

Solution to part (a): We determine the net magnetic field at point P_1 as the sum of two contributions: the magnetic field at P_1 due to wire 1 and the magnetic field at P_1 due to wire 2. Even if we were not asked for the direction of the net magnetic field, the direction of each of the two contributions must be determined since we need to know whether the two contributions should be added or subtracted from each other.

The contribution at point P_1 due to wire 1 is calculated with $d_1 = l_1$:

$$|\boldsymbol{B}_1|(P_1) = \frac{8.0 \times 10^{-7}\ \mathrm{T \cdot m}}{0.06\ \mathrm{m}} = 1.33 \times 10^{-5}\ \mathrm{T} \quad \textbf{(4)}$$

This contribution is directed out of the paper in Fig. 20.42 due to the right hand rule. The contribution at the same point due to wire 2 is calculated with $d_2 = l_1 + l_2$:

$$|\boldsymbol{B}_2|(P_1) = \frac{6.0 \times 10^{-7}\ \mathrm{T \cdot m}}{0.11\ \mathrm{m}} = 5.5 \times 10^{-6}\ \mathrm{T} \quad \textbf{(5)}$$

This contribution is directed into the plane of the paper in Fig. 20.42. Therefore, the net magnetic field at P_1 is $|\mathbf{B}|(P_1) = 7.8 \times 10^{-6}$ T, pointing out of the plane of the paper in Fig. 20.42.

Solution to part (b): We determine the net magnetic field at point P_2 in the same fashion as in part (a). We find with $d_1 = l_2 + l_3$:

$$|\boldsymbol{B}_1|(P_2) = 8.0 \times 10^{-6}\ \mathrm{T} \quad \textbf{(6)}$$

This contribution is directed into the plane of the paper in Fig. 20.42. Also with $d_2 = l_3$:

$$|\boldsymbol{B}_2|(P_2) = 1.20 \times 10^{-5}\ \mathrm{T} \quad \textbf{(7)}$$

This contribution is directed out of the paper in Fig. 20.42. Therefore, the net magnetic field at P_2 is $|\mathbf{B}|(P_2) = 4.0 \times 10^{-6}$ T pointing out of the plane of the paper in Fig. 20.42.

Solution to part (c): From parts (a) and (b) we know that the contributions due to the two wires have to be subtracted from each other for all points to the left or to the right of the arrangement. Thus, on either side there could be a point where the net magnetic field is indeed vanishing.

To set the problem properly up for this, we have to define two conventions:

- We assign a positive sign to the magnetic field value at a point where that field points out of the plane of paper in Fig. 20.42, and we assign a negative sign if the magnetic field points into the plane of the paper.
- We define an x–axis perpendicular to the two wires in the plane of Fig. 20.42. The positive x–axis points toward the right and the origin ($x = 0$) is chosen at the position of wire 1.

With these conventions, we combine the two equations for the magnitudes of the magnetic field for wire 1 and wire 2, i.e., Eqs. [2] and [3]:

$$|\boldsymbol{B}|(x) = \frac{-8 \times 10^{-7}\ \mathrm{T \cdot m}}{x} + \frac{6 \times 10^{-7}\ \mathrm{T \cdot m}}{x - l_2} \quad \textbf{(8)}$$

We find the x–position at which the magnetic field becomes zero by setting Eq. [8] equal to zero:

$$\frac{-8 \times 10^{-7}(x - l_2) + 6 \times 10^{-7} x}{x \cdot (x - l_2)} = 0 \quad \textbf{(9)}$$

This simplifies to:

$$x = +4 \cdot l_2 \quad \textbf{(10)}$$

Thus, the point with zero magnetic field lies 20 cm to the right of wire 1 or 15 cm to the right of wire 2.

Problem 20.3

A conducting wire has a mass of 10 g per metre of length. The wire carries a current of 20 A and is suspended directly above a second wire of the same type that carries a current of 35 A. How far do you have to close the separation distance between the wires so that the upper wire is balanced at rest by magnetic repulsion?

Solution: To establish mechanical equilibrium, the magnetic force per unit length of the wire, acting on the top wire, must be equal to the weight per unit length of the wire. Thus, the free–body diagram of the system (e.g. choosing 1 metre as the unit length of the wire) contains two vertical force vectors, the weight downward and the magnetic force upward.

The weight per unit length is labelled $|\mathbf{W}|/L$ and its value for the wire in the present problem is given by:

$$\frac{|\boldsymbol{W}|}{L} = \frac{m \cdot g}{L} = \frac{10\ \mathrm{g}}{1\ \mathrm{m}}\left(9.8\ \frac{\mathrm{m}}{\mathrm{s}^2}\right) = 0.098\ \frac{\mathrm{N}}{\mathrm{m}} \quad \textbf{(11)}$$

Now we use Eq. [20.2] for the magnetic force per unit length of a wire:

$$\frac{|\boldsymbol{F}_{mag}|}{l} \propto \frac{I_1 \cdot I_2}{d} \quad \textbf{(12)}$$

Thus, we require:

$$\frac{\mu_0 \cdot I_1 \cdot I_2}{2 \cdot \pi \cdot d} = \frac{|\boldsymbol{W}|}{L} \quad \textbf{(13)}$$

Isolating the unknown variable d in Eq. [13], we find:

$$d = \frac{\mu_0 \cdot I_1 \cdot I_2}{2 \cdot \pi \, (|\mathbf{W}|/L)} \quad \textbf{(14)}$$

This yields:

$$d = \frac{\left(1.26 \times 10^{-6} \, \frac{\text{N}}{\text{A}^2}\right)(20 \text{ A})(35 \text{ A})}{2 \cdot \pi \left(0.098 \, \frac{\text{N}}{\text{m}}\right)} = 1.4 \text{ mm} \quad \textbf{(15)}$$

The distance between the wires is 1.4 mm.

Problem 20.5

Fig. 20.43 shows two parallel wires that carry currents $I_1 = 100$ A and I_2. The top wire is held in position; the bottom wire is prevented from moving sideways but can slide up and down without friction. If the wires have a mass of 10 g per metre of length, calculate current I_2 such that the lower wire levitates at a position 4 cm below the top wire.

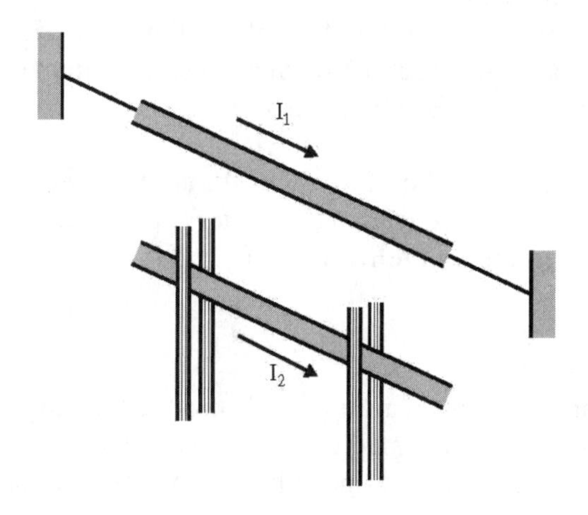

FIGURE 20.43

Solution: The magnetic force exerted on the lower wire by the upper wire is directed upward since their currents flow in the same direction. In mechanical equilibrium, we write for a piece of the lower wire of length L:

$$\sum_i F_{i,\,y} = 0 = \frac{\mu_0 \cdot I_1 \cdot I_2 \cdot L}{2 \cdot \pi \cdot d} - m \cdot g \quad \textbf{(16)}$$

in which the upward force component is the magnetic force taken from Eq. [13] and the downward force is the weight. d is the distance between the two wires in mechanical equilibrium, and m is the mass of a wire segment of length L. We determine the current in the lower wire, I_2, from Eq. [16]:

$$I_2 = \frac{2 \cdot \pi \cdot d}{\mu_0 \cdot I_1} \cdot \frac{m}{L} \cdot g$$

$$= \frac{2 \cdot \pi \, (0.04 \text{ m})\left(0.01 \, \frac{\text{kg}}{\text{m}}\right)\left(9.8 \, \frac{\text{m}}{\text{s}^2}\right)}{\left(1.26 \times 10^{-6} \, \frac{\text{N}}{\text{A}^2}\right)(100 \text{ A})} \quad \textbf{(17)}$$

This yields:

$$I_2 = 195 \text{ A} \quad \textbf{(18)}$$

Problem 20.7

A conductor carries a current of 10 A in a direction that makes a 30^0 angle with the magnetic field of strength $B = 0.3$ T. What is the magnitude of the magnetic force on a 5-m segment of the conductor?

Solution: We use the magnetic force on a length l of conductor, Eq. [20.5], but since we have been asked to find the magnitude of the magnetic force we can use Eq. [20.8] where the given quantities are the magnitudes of the relevant vector quantities and the angle θ is the angle between the magnetic field direction and the direction of the current. We can now find the magnetic force as $F_{mag} = 7.5$ N.

Problem 20.9

Fig. 20.45 shows the cross–sections of four long, parallel, current–carrying conductors. Each current is 4.0 A, and the distance between neighbouring conductors is $L = 0.2$ m. A dot on the conductor means the current is

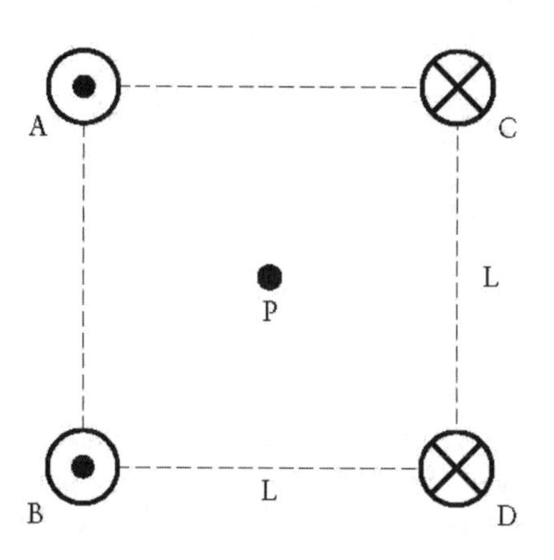

FIGURE 20.45

flowing out of the plane of the paper and a cross means it flows into the plane of the paper. Calculate the magnitude and determine the direction of the magnetic field at point P at the centre of the square shown.

Solution: First of all we must consider the directions of the fields created by each conductor at the location of point P. Using the right–hand rule where the thumb points in the direction of current and the field curls in the direction of the fingers, the directions of each magnetic field contribution are: B_A up and to the right, B_B up and to the left, B_C up and to the left and B_D up and to the right. Note that all angles are at 45^0 above horizontal.

The magnitude of each magnetic field is equal since the currents are all equal and the distances between each conductor and point P are the same. This distance may be found as the hypotenuse of the triangle that makes up a quarter of the square configuration: $d = 0.141$ m. Each magnetic field is then:

$$B = \frac{\mu_0 \cdot I}{2 \cdot \pi \cdot d} = \frac{1.26 \times 10^{-6} \frac{\text{N}}{\text{A}^2}}{2\pi} \frac{4.0 \text{ A}}{0.141 \text{ m}} \quad (19)$$

$$= 5.69 \times 10^{-6} \text{ T}$$

The total magnetic field in each direction is then twice this value, once directed up and right, and once directed up and left, as shown in Fig. 20.47. The total magnetic field may be calculated using the Pythagorean theorem as $B_P = 1.6 \times 10^{-5}$ T. This field is directed upward because the left and right component cancel, leaving us with only upward components from each contribution.

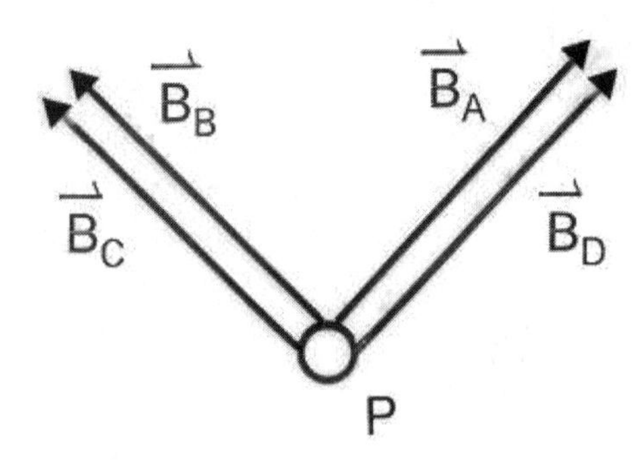

Figure 20.47

Problem 20.11

The index of refraction of a slab of glass is $n = 1.52$.
(a) What is Brewster's angle if the glass is located in air?
(b) Modify the formulas in the text such that you can calculate Brewster's angle if the glass slab is in water. What is its value?

Solution to part (a): If the glass slab is in air, we can use for Brewster's angle

$$n = \tan(\theta_p) \quad \textbf{(20)}$$

so that $\theta_p = 56.7^0$.

Solution to part (b): If the glass slab is in water then the equation for Brewster's angle is modified. The derivation comes from the law of refraction. If n_1 equals something other than 1 (for air), we have the condition:

$$\frac{n_2}{n_1} = \frac{\sin(\theta_p)}{\sin(\theta_2)} = \frac{\sin(\theta_p)}{\sin(90^0 - \theta_p)} = \frac{\sin(\theta_p)}{\cos(\theta_p)} = \tan(\theta_p) \quad \textbf{(21)}$$

For glass in water we find $\tan\theta_p = 1.52/1.33 = 1.14$, i.e., $\theta_p = 48.7^0$.

Problem 20.13

A light ray is incident on flint glass with $n = 1.65$. Calculate the angle of refraction of a transmitted light ray if the reflected ray is fully polarized.

Solution: We use Eq. [20]. Since the angle of refraction and the angle of reflection (Brewster's angle) must sum to 90^0, the angle of refraction is $\theta_{refracted} = 31.2^0$.

Problem 20.15

If light is incident at an angle θ from a medium of index of refraction n_1 to a medium with n_2 such that the angle between the reflected and refracted beams is β, show that:

$$\tan\theta = \frac{n_2 \cdot \sin\beta}{n_1 - n_2 \cdot \cos\beta} \quad \textbf{(22)}$$

Hint: Use the formula for $\sin(\alpha_1 + \alpha_2)$ from the Math Review *Trigonometry* at the end of Chapter 17.

Solution: Start with the law of refraction.

Problem 20.17

Light of wavelength λ_0 in vacuum has a wavelength of $\lambda_w = 438$ nm in water and a wavelength of $\lambda_b = 390$ nm in benzene.
(a) What is the wavelength λ_0 in a vacuum?
(b) Using only the given information, determine the ratio

of the index of refraction of benzene to that of water.

Solution to part (a): To relate the speed of light in a medium to the medium's index of refraction, we combine the equation that connects the speed of light across an interface and Snell's law. We did this before, e.g., in Example 18.5, where we found in Eq. [18.25] the relation: $c/v_{medium} = n_{medium}$ with c the speed of light in a vacuum. For the current problem, we want to connect this formula to the wavelength and frequency values of light of a specific colour:

- for the light in water (index w): $v_w = \lambda_w \cdot f$, and
- for the light in a vacuum (index 0): $c = \lambda_0 \cdot f$.

Note that we do not use an index with the frequency since the light frequency does not change across an interface. Substituting these terms in $c/v_{medium} = n_{medium}$, we get:

$$n_w = \frac{\lambda_0 \cdot f}{\lambda_w \cdot f} \quad \textbf{(23)}$$

which yields:

$$\lambda_0 = n_w \cdot \lambda_w = 1.33\,(438\text{ nm}) = 583\text{ nm} \quad \textbf{(24)}$$

Solution to part (b): For the comparison with benzene we use Eq. [23] twice to calculate the ratio of the two index of refraction:

$$\frac{n_b}{n_w} = \frac{\dfrac{\lambda_0 \cdot f}{\lambda_b \cdot f}}{\dfrac{\lambda_0 \cdot f}{\lambda_w \cdot f}} = \frac{\lambda_w}{\lambda_b} = \frac{438\text{ nm}}{390\text{ nm}} = 1.12 \quad \textbf{(25)}$$

Problem 20.19

The index of refraction of red light in water is $n = 1.331$, and for blue light it is $n = 1.340$. If a ray of white light enters the water at an angle of incidence of 83^0, what are the underwater angles of refraction for the two light components?

Solution: We use the law of refraction for each of the indices of refraction yielding: $\theta_{red} = 48.2^0$ and $\theta_{blue} = 47.8^0$.

Problem 20.21

A typical human skin temperature is 35^0C. At what wavelength does the radiation emitted by the human body reach its peak?

Solution: We use Wien's displacement law in Eq. [20.24] to find the wavelength at which the radiation from a body at a given temperature has its peak. Note that this equation requires a temperature in Kelvin. We find $\lambda_{max} = 941\ \mu$m. This wavelength is in the infrared part of the electromagnetic spectrum.

CHAPTER TWENTY–ONE

The human body in outer space: Circular motion and a first look at radiation

MULTIPLE CHOICE AND CONCEPTUAL QUESTIONS

Question 21.1

Fig. 21.1 contains four axes for the temperature, height, pressure and $F_{gravity}/m$ as a multiple of g.
(a) Which axis is linear?
(b) Which axis is logarithmic?

Answer to part (a): The temperature axis is linear.

Answer to part (b): The axis for the height is logarithmic.

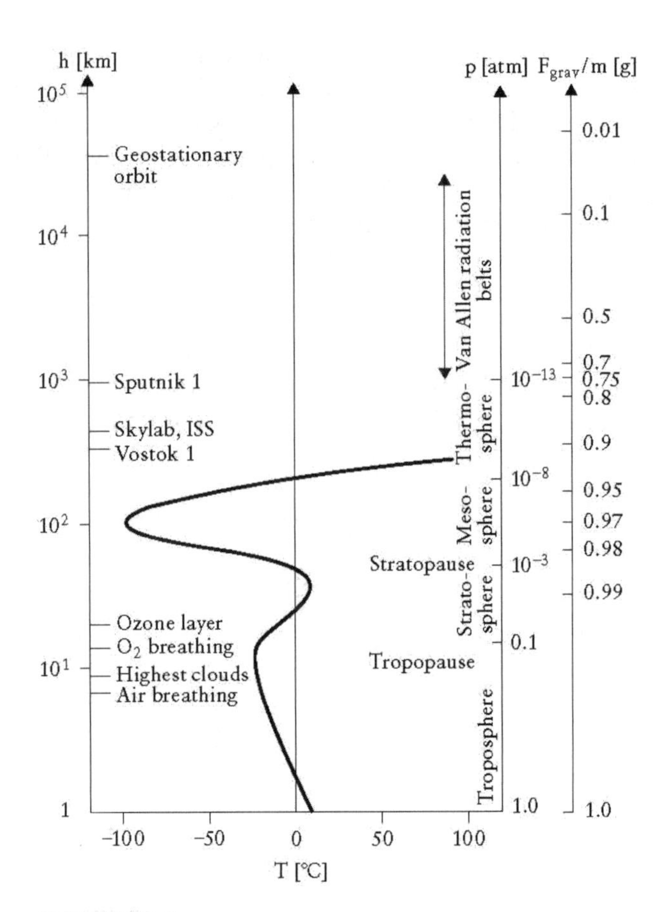

FIGURE 21.1

Question 21.3

What happens to a gas–filled balloon that rises high in the atmosphere, such as a weather balloon? Will it collapse or expand? Will it leave the atmosphere or come to rest?

Answer: It will expand in an isothermal atmosphere. It will expand in the actual atmosphere of Fig. 21.1 because the temperature varies initially less than the pressure. The balloon rises based on buoyancy. In outer space there is too little gas to displace, thus the balloon reaches a maximum height (at which it better not yet rupture unless a parachute is included to prevent it from crashing back to Earth's surface).

Question 21.5

(a) Can the space shuttle be located stationary above its mission control centre in Texas?
(b) Can a European TV satellite be located above a ground station in Athens, Greece?

Answer to part (a): No; the space shuttle doesn't operate in a geostationary orbit.

Answer to part (b): No. Even in the geostationary orbit the satellite must be located above a position on the equator. Athens, Greece, is located at 38^0 N; about the same northern latitude as San Francisco in North America, or Seoul in Asia.

Question 21.7

The mass of Mars is $m_{Mars} = 0.102 \cdot m_{Earth}$. What gravitational acceleration g_{Mars} as a multiple of g_{Earth} affects an astronaut on Mars? *Note*: R_{Mars} = 3400 km and R_{Earth} = 6370 km.

Answer: The acceleration due to gravity at the surface of a planet is given in Eq. [21.3] where the gravitational constant is $G^* = 6.67 \times 10^{-11}\ N \cdot m^2/kg^2$. The ratio of the acceleration due to gravity on Mars relative to that one Earth will be:

$$\frac{g_{Mars}}{g_{Earth}} = \frac{m_{Mars}}{m_{Earth}} \cdot \frac{R_{Earth}^2}{R_{Mars}^2}$$

$$= 0.102\left(\frac{6370\text{ km}}{3400\text{ km}}\right)^2 = 0.358 \quad (1)$$

The acceleration due to gravity on Mars is approximately 36% that on Earth.

Question 21.9
An object moves in uniform circular motion when a constant force acts perpendicular to its velocity vector. What happens if the force is not perpendicular?

Answer: Its component along the path of the object accelerates the object; i.e., its motion is no longer uniform. The perpendicular component causes a curved path.

Question 21.11
An object moves with uniform circular motion.
(a) Is its velocity constant?
(b) Is its speed constant?
(c) Is its acceleration constant?

Answer to part (a): No; its x– and y–components vary continuously, i.e., the velocity in a Cartesian coordinate system is not constant.

Answer to part (b): Yes; its speed along the path is constant as no acceleration acts in that direction for a uniform motion.

Answer to part (c): Yes if we refer to its magnitude, no if we refer to acceleration as a vector in a Cartesian coordinate system.

Question 21.13
A bucket of water can be whirled in a vertical loop without any water being spilled. Why does the water not flow out when the bucket is at the top of the loop?

Answer: The gravitational acceleration downward is smaller than the centripetal acceleration required to hold it on a circular path; as a result the bottom of the bucket has to push the water down when it is at the top of the loop. Newton's third law then states that the water pushes the bottom of the bucket upward. Compare the water to a person on a fast roller coaster, as opposed to a slow Ferris wheel.

Question 21.15
What happens to a mobile but initially stationary point charge if a magnet travels past its location?

Answer: The same that would happen if the point charge were moving with the opposite velocity vector past the stationary magnet. For example, this case occurs in Faraday's experiment when a magnet moves through a conductor loop.

Question 21.17
In Fig. 21.19, the bar magnet within Earth is shown with its south pole underneath northern Canada. Is this correct?

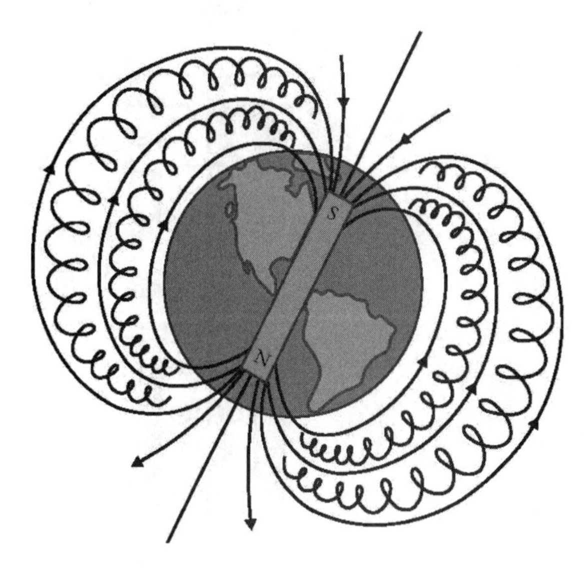

FIGURE 21.19

Answer: Yes, for the Earth's magnetic North Pole to be near the geographic North Pole, the internal bar magnet must have its south pole beneath the northern hemisphere. Note that it is the North pole of a bar magnet inside a magnetic compass that is attracted to the Earth's magnetic pole located in northern Canada.

Question 21.19
Two charged particles travel into a region in which a magnetic field acts perpendicular to the particles' velocity vectors. What do you conclude if they are deflected in opposite directions?

Answer: They carry opposite charges.

Question 21.21
Why do cosmic–ray particles strike Earth more often

near the poles than near the equator?

Answer: Due to the shape of the magnetic field of Earth (see Fig. 21.19). The charged particles spiral about the magnetic field lines that in turn intersect with the surface of Earth near the magnetic poles.

Question 21.23
What happens in Fig. 21.20 if the three slits S_1 to S_3 are opened to twice their width?

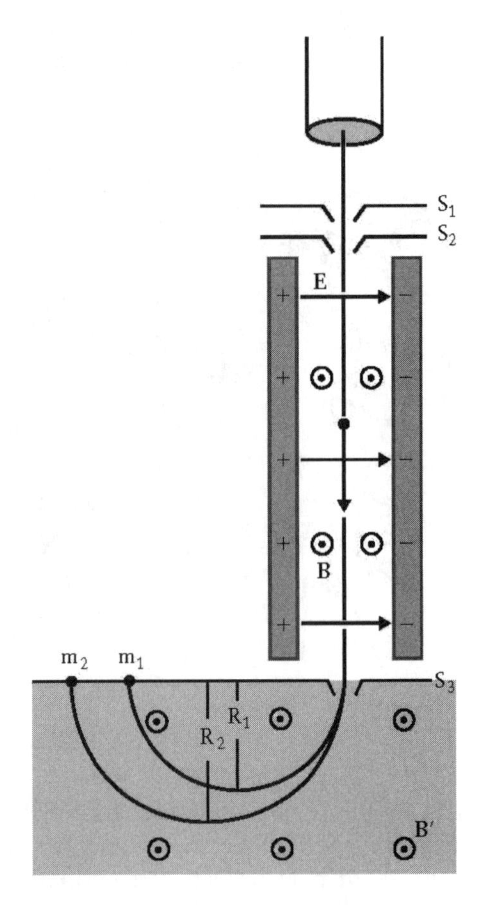

FIGURE 21.20

Answer: Particles enter the Wien filter with a larger range of angles. This allows a broader range of velocities to pass the Wien filter, increasing the intensity, but decreasing the precision of the selected speed. This leads to the detection of a broadened peak in the sector magnet. Sharpness of mass lines and their intensity have to be optimized for any actual instrument; the sharpest mass peaks follow when the slitwidths are driven toward zero, but the peak intensity is concurrently driven below the detection limit of the instrument.

ANALYTICAL PROBLEMS

Problem 21.1
A roller coaster ride includes a circular loop with radius $R = 10$ m.
(a) What minimum speed must the car have at the top to stay in contact with the tracks?
(b) What minimum speed must the car have when entering the loop to satisfy the solution in part (a)?

Solution to part (a): The system of interest is the car passing through the roller coaster loop. The circular track is part of its environment. Two forces act on the car: its weight and the contact force exerted by the track. Of these two forces, only the weight is not zero at the top point when the car passes with the least possible speed.

The free–body diagram for this problem is trivial with a single force acting on the system straight down. We use the coordinate system in which the plane of the circular motion is identified by two coordinates attached to the system, labelled ∥ parallel to the velocity vector and ⊥ perpendicular to the velocity vector. The direction perpendicular to this plane is the z–direction. In this coordinate system, Newton's laws are applied in the form of Eq. [21.18]:

$$\begin{aligned} &(I) \quad \sum_i F_{i,\perp} = m\frac{v^2}{r} \\ &(II) \quad \sum_i F_{i,\parallel} = 0 \\ &(III) \quad \sum_i F_{i,z} = 0 \end{aligned} \qquad \textbf{(2)}$$

Since there is no motion in the z–direction, the respective equation can be neglected. Also, when the car is at the top of the loop, no force acts parallel to the velocity vector, i.e., tangential to the track. The only remaining equation of interest is the first formula in Eq. [2]:

$$\sum_i F_{i,\perp} = -m \cdot g = -m\frac{v_{top}^2}{R} \qquad \textbf{(3)}$$

in which R is the radius of the loop, m is the mass of the car and v_{top} is the minimum speed. Note that at higher speeds an additional force has to be added to the left hand side of Eq. [3] which is due to the normal force exerted by the track onto the car.

Eq. [3] is solved by separating the minimum speed v_{top}:

$$v_{top} = \sqrt{g \cdot R} = \sqrt{\left(9.8\ \frac{\mathrm{m}}{\mathrm{s}^2}\right)(10\ \mathrm{m})} = 9.9\ \frac{\mathrm{m}}{\mathrm{s}} \quad \mathbf{(4)}$$

Note that this result is independent of the mass of the car.

Solution to part (b): We use the conservation of energy. To do so, we compare two states for the system:
- when the car is at the top of the loop, at which point we know most details about it, and
- when the car is at the bottom of the loop, because we are asked about the speed of the car at that point:

$$\begin{aligned} E_{total,\ top} &= E_{kin,\ top} + E_{pot,\ top} = \\ E_{total,\ bottom} &= E_{kin,\ bottom} + E_{pot,\ bottom} \end{aligned} \quad \mathbf{(5)}$$

Remember that we can choose the reference height $y = 0$ freely when writing the conservation of energy in the form of Eq. [5]. In the current case, choosing the origin for the height, $y = 0$, at the bottom of the loop simplifies the calculation:

$$\frac{1}{2} m \cdot v_{top}^2 + m \cdot g\,(2 \cdot R) = \frac{1}{2} m \cdot v_{bottom}^2 \quad \mathbf{(6)}$$

For the second term on the left hand side of Eq. [6] we used the fact that the highest point of the loop is twice the radius above the lowest point.

We substitute the result from part (a) for v_{top} on the left side of Eq. [6]:

$$\frac{1}{2} m \cdot g \cdot R + m \cdot g\,(2 \cdot R) = \frac{1}{2} m \cdot v_{bottom}^2 \quad \mathbf{(7)}$$

which yields:

$$v_{bottom} = \sqrt{5 \cdot g \cdot R} \quad \mathbf{(8)}$$

With the given value of R, we calculate $v_{bottom} = 22$ m/s, which corresponds to 80 km/h.

Problem 21.3
Calculate the orbital speed for the two Russian missions shown in Fig. 21.1. Use the mass and the radius of Earth from the examples in this chapter.

Solution: Eq. [21.41] applies to any orbiting satellite, not just the ISS:

$$v_{ISS} = \sqrt{G^* \frac{M_{Earth}}{R_{Earth} + h}} \quad \mathbf{(9)}$$

The height above ground for the two Russian missions shown in Fig. 21.1, Sputnik 1 (first unmanned satellite (83 kg, 58 cm diameter), launched in October 1957) and Vostok 1 (carrying Yury Gagarin into space in April 1961) can be read off the figure. Note that its scale is logarithmic: $h_{Sputnik} = 946$ km and $h_{Vostok} = 327$ km. (The complete name of the Sputnik 1 mission was "Iskustvennyi Sputnik Zemli" which translates as "fellow world traveller of Earth.")
- We find for Sputnik 1:

$$v_{Sputnik} = \sqrt{G^* \frac{M_{Earth}}{(R_{Earth} + h_{Sputnik})}}$$

$$= \sqrt{\frac{\left(6.67 \times 10^{-11}\ \frac{\mathrm{m}^3}{\mathrm{kg} \cdot \mathrm{s}^2}\right)(5.98 \times 10^{24}\ \mathrm{kg})}{6.37 \times 10^6\ \mathrm{m} + 9.46 \times 10^5\ \mathrm{m}}} \quad \mathbf{(10)}$$

$$= 7385\ \frac{\mathrm{m}}{\mathrm{s}}$$

- We further find for Vostok 1 $v_{Vostok} = 7715$ m/s. Comparing these two speeds with the speed of the ISS at 7660 m/s from Example 21.6, we notice that the absolute speeds vary little due to the relatively small variations of the spacecraft's distance from the centre of Earth.

Problem 21.5
Eq. [21.40] also applies to the circular motion of a planet around the Sun:

$$G^* \frac{M_{Earth} \cdot m_{ISS}}{(R_{Earth} + h)^2} = m_{ISS} \frac{v_{ISS}^2}{(R_{Earth} + h)} \quad \mathbf{(11)}$$

Derive from this equation Kepler's third law, which states that $T^2 \propto r^3$ if T is the planet's period and r its distance from the Sun.

Solution: For the planet/Sun system Eq. [11] reads:

$$G^* \frac{M_{Sun} \cdot m_{planet}}{r^2} = m_{planet} \frac{v_{planet}^2}{r} \quad \mathbf{(12)}$$

in which m_{planet} and a factor r can be cancelled on both sides. With $v_{planet} = 2 \cdot \pi \cdot r/T$ follows:

$$\frac{r^3}{T^2} = \frac{G^* \cdot M_{Sun}}{4 \cdot \pi^2} \quad \textbf{(13)}$$

This is Kepler's third law because all terms on the right hand side are constants in our solar system.

Problem 21.7
An airplane flies in a horizontal circle with a speed of 100 m/s. The pilot is a standard man (see Table 3.3). The maximum acceleration of the pilot should not exceed $7 \cdot g$, with g the gravitational acceleration. What is the minimum radius of the airplane's circular path?

Solution: If the maximum acceleration a pilot may experience during a horizontal circle is seven times the gravitational acceleration, then we are concerned with the pilot's centripetal acceleration. Therefore we will find the conditions for the centripetal acceleration to equal $7 \cdot g$ and take those conditions as the limits for what the standard man may do.

Our condition is then for the circular motion that $a_\perp = 7 \cdot g$. Since the airplane travels at 100 m/s, we can find the radius of the circle as $r = 146$ m.

This is the minimum radius of the airplane's circular path; any smaller and the centripetal acceleration will surpass the given limit. Note: this result is independent of the pilot's mass.

Problem 21.9
A child of mass $m = 40$ kg takes a ride on a Ferris wheel that has a radius of 9.0 m and spins four times per minute.
(a) What is the child's centripetal acceleration?
(b) What is the force the seat exerts on the child at the highest and lowest points, and when the child is halfway between?

Solution to part (a): To find the centripetal acceleration of the child we need to have the radius of the Ferris wheel (which is given) and the speed of the Ferris wheel, which we will need to determine. The period of the ride is given in the information "spins four times per minute". This means one period takes 15 s. The speed of the ride is then the total distance travelled in one period (the circumference) divided by that time, $v = 2 \cdot \pi \cdot r/t$ which yields 3.77 m/s. The centripetal acceleration is then: $a = 1.6$ m/s².

Solution to part (b): At the top, gravity exerts a force downward equal to the child's weight, $W = 390$ N, while the seat exerts a normal force upwards. Together these forces must sum like vectors to create a net force which causes the centripetal acceleration found in part (a). The centripetal acceleration will point downward because the child is accelerated toward the centre of the circular path, which is downward because the child is at the top of the circle. This means:

$$F_{net} = -W + N_{top} = -m \cdot a_\perp \quad \textbf{(14)}$$

which yields:

$$N_{top} = 390\text{ N} - 64\text{ N} = 330\text{ N} \quad \textbf{(15)}$$

At the bottom, gravity exerts a force downward equal to the child's weight (still 390 N) while the seat exerts a normal force upward. Together these forces must sum like vectors to create a net force which causes the centripetal acceleration found in part (a). The centripetal acceleration will in this case point *upward* because the child is accelerated toward the centre of the circular path, which is upward at the bottom of the circle. This means:

$$F_{net} = -W + N_{bottom} = +m \cdot a_\perp \quad \textbf{(16)}$$

which yields:

$$N_{top} = 390\text{ N} + 64\text{ N} = 450\text{ N} \quad \textbf{(17)}$$

Halfway between the top and the bottom, gravity will still exert a force of 390 N downward and since the child is moving with constant speed the seat must exert a normal force which balances the child's weight. The normal force must also have a component directed horizontally since the centre of the circular path is directly horizontal from the child. No component of the weight can exert a force toward the centre so the normal force must provide the entire centripetal acceleration, a horizontal component of 64 N acts toward the centre of Ferris wheel.

Problem 21.11
A long, straight wire in a vacuum system carries a current of 1.5 A. A low–density, 20-eV electron beam is directed parallel to the wire at a distance of 0.5 cm. The electron beam travels against the direction of the current in the wire. Find
(a) the magnitude of the magnetic force acting on the electrons in the electron beam, and
(b) the direction in which the electrons are deflected from their initial direction.

Figure 21.25 ⇒

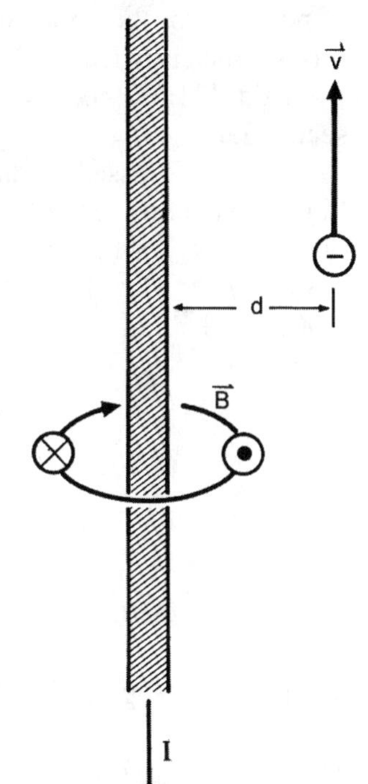

Solution to part (a): The problem is illustrated in Fig. 21.25. The figure indicates the direction of the magnetic field **B** due to the current in the wire. As shown, the magnetic field at the position of the electron beam points out of the plane of the paper. We first use Eq. [20.3]:

$$|\boldsymbol{B}| = \frac{\mu_0}{2 \cdot \pi} \cdot \frac{I}{d} \quad \textbf{(18)}$$

and Eq. [21.54]

$$f_{mag} = q \cdot v \cdot B \quad \textbf{(19)}$$

to determine the magnitude of the force acting on each electron. Eq. [18] allows us to calculate the magnitude of the magnetic field of the current at a given distance d from the wire:

$$B = \frac{\left(1.26 \times 10^{-6} \frac{\text{N}}{\text{A}^2}\right) 1.5 \text{ A}}{2 \cdot \pi \,(5 \times 10^{-3} \text{ m})} = 6 \times 10^{-5} \text{ T} \quad \textbf{(20)}$$

With this result, we use Eq. [19] to determine the magnitude of the magnetic force on each electron in the beam. Eq. [19] contains the speed of the electrons which we calculate from the given kinetic energy of 20 eV:

$$E_{kin} = 20 \text{ eV} \left(1.6 \times 10^{-19} \frac{\text{J}}{\text{eV}}\right) = 3.2 \times 10^{-18} \text{ J} = \frac{1}{2}\, m_{electron} \cdot v^2 \quad \textbf{(21)}$$

which leads to $v = 2.65 \times 10^6$ m/s when the mass of the electron is taken from Table 13.1. Note that it is always useful to check such a speed, calculated from an energy for an atomic or sub–atomic particle, against the speed of light. In the current case we find that the speed is just below 1% of the speed of light; thus, our classical calculation (using $E_{\text{kin}} = ½ \cdot m \cdot v^2$) is valid. If the speed exceeds a few percent of the speed of light a relativistic correction would have to be used.

The values from Eqs. [20] and [21] are substituted into Eq. [19]:

$$f_{mag} = (1.6 \times 10^{-19} \text{ C}) \left(2650 \frac{\text{km}}{\text{s}}\right) (6 \times 10^{-5} \text{ T}) = 2.5 \times 10^{-17} \text{ N} \quad \textbf{(22)}$$

Solution to part (b): The direction of the force is determined with the right hand rule as illustrated in Fig. 21.26 for both positively and negatively charged particles. We apply Fig. 21.26(b) to the directions shown in the sketch of Fig. 21.25. With the velocity vector directed upward and the magnetic field at the position of the electron pointing out of the plane of the paper, the force is directed toward the wire. Thus, the electron is attracted toward the wire.

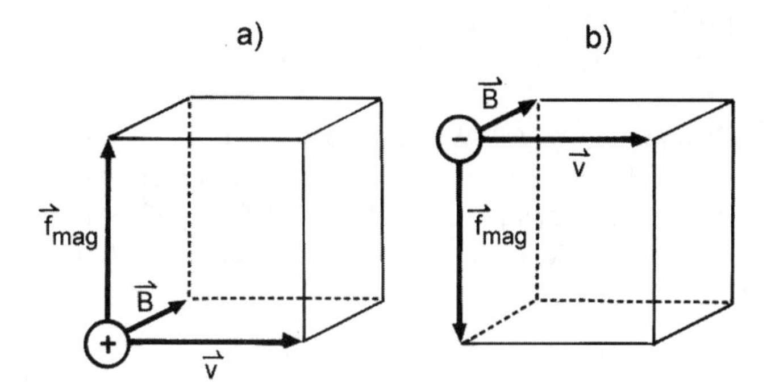

Figure 21.26

Problem 21.13

Sodium ions Na^+ move at 0.5 m/s through a blood vessel. The blood vessel is in a magnetic field of $B = 1.0$ T. The blood flow direction subtends an angle of 45^0 with the magnetic field. What is the magnetic force on the blood vessel due to sodium ions if the blood vessel contains 0.1 L blood with a sodium concentration $c = 70$ mmol/L?

Solution: This problem requires us to first calculate the number of sodium ions and then find the magnetic force acting on all the ions at once since they travel together in the blood vessel. The number of sodium ions is found from the volume, the molar concentration and Avogadro's number as $N_{\text{Na+}} = 4.2 \times 10^{21}$ ions.

The magnetic force on the blood vessel is the total force on all the ions. We can find the magnitude of this force from Eq. [21.56]: $F_{\text{net}} = 240$ N.

Problem 21.15

A proton moves in uniform circular motion perpendicular to a uniform magnetic field with $B = 0.8$ T. What is the period of its motion?

Solution: The radius of the proton's circular path is given in Eq. [21.61]. From Eq. [21.62] we see that the period of this circular motion is related to the speed and radius by $v/r = 2 \cdot \pi/T$.

With this is mind we rearrange the equation for the radius of the path in the form:

$$\frac{v}{r} = \frac{q \cdot B}{m} = \frac{2 \cdot \pi}{T} \quad \textbf{(23)}$$

which we re–arrange for the period: $T = 8.2 \times 10^{-8}$ s.

Problem 21.17

A mass spectrometer is used to separate isotopes. If the beam emerges with a speed of 250 km/s and the magnetic field in the mass selector is 2 T, what is the distance between the collectors for

(a) ^{235}U and ^{238}U, and

(b) ^{12}C and ^{14}C?

Solution to part (a): We need not to quantify the properties of the Wien filter in this case because the speed of the selected ions is given. We quantify Eq. [21.61] for each isotope separately and then calculate the difference in position for the collector. Note from Fig. 21.20 that this distance is twice the difference in radii as calculated from Eq. [21.61] because both isotopes enter the 180^0 sector field.

As discussed in Chapter 23, the isotope labels allow us to determine the mass of each isotope: ^{235}U has a mass of $235 \cdot u$ in which u is the atomic unit ($1\ u = 1.6605677 \times 10^{-27}$ kg). Therefore, the mass of ^{235}U is 3.902×10^{-25} kg. ^{238}U has a mass of $238 \cdot u$, which is 3.952×10^{-25} kg. This leads to:

$$r_{U-235} = \frac{(3.902 \times 10^{-25}\ kg)\left(2.5 \times 10^{5}\ \frac{m}{s}\right)}{(1.6 \times 10^{-19}\ C)\,(2.0\ T)} = 0.305\ m \quad \textbf{(24)}$$

and $r_{U-238} = 0.309$ m. Thus, their separation is 8 mm.

Solution to part (b): The analogous calculation for the two carbon isotopes leads to $r_{C-12} = 0.0156$ m and $r_{C-14} = 0.0182$ m. i.e., a separation of 5 mm. Note that a smaller magnetic field leads to a larger separation.

CHAPTER TWENTY–TWO

Chemical bonds and X–rays: Atomic and molecular physics

MULTIPLE CHOICE AND CONCEPTUAL QUESTIONS

Question 22.1

Once a helicopter takes off from its landing platform, one would expect that the non–zero friction between the rotor axis and its bearings causes the helicopter cabin to spin out of control. Look at a photograph of a helicopter and determine how this effect is prevented.

Answer: Smaller helicopters have a second, smaller rotor turned 90^0 at the tail. Large helicopters have two rotors that spin in opposite directions.

Question 22.3

Why do figure skaters pull their arms close to their chest when they want to spin fast, e.g., during a fast spin or a triple jump?

Answer: We defined the angular momentum L in Eq. [22.14]:

$$L = r \cdot m \cdot v \cdot \sin\varphi \qquad (1)$$

The angular momentum is conserved during a jump. To achieve a maximum speed of rotation v, the mass distribution of the rotating object must be as close as possible to the axis of rotation; i.e., the product $m \cdot r$ has to be a minimum.

Question 22.5

What is wrong with Rutherford's atomic model (see Fig. 22.4), which is based on an analogy to the planetary system? (A) The planetary system has eight planets, but the atom may have more or fewer electrons. (B) The planetary system formed after a supernova ejected large amounts of gas/dust into space. No such event precedes the formation of an atom. (C) The planets are not charged electrically like the electron. (D) An electron orbiting a positive charge must lose energy via electromagnetic radiation. (E) The International Astronomic Society can decide how many objects in the solar system qualify as planets, but it has no jurisdiction to decide how many electrons are present in the shell of an atom.

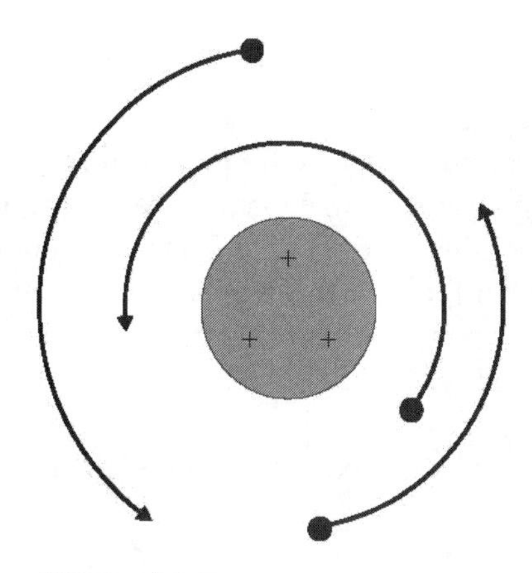

FIGURE 22.4

Answer: (D)

Question 22.7

Bohr postulated that an electron in a particular orbit does not radiate. What consequence does this assumption have? (A) The linear momentum of the electron is conserved. (B) The total energy of the electron is conserved. (C) The kinetic energy of the electron is conserved. (D) The angular momentum of the electron varies only with time.

Answer: (B). Otherwise it would lose energy continuously (i.e., it would be a closed system and not an isolated system).

Question 22.9

De Broglie's wavelength of a subatomic particle depends on its kinetic energy E as: (A) $\lambda \propto E$, (B) $\lambda \propto E^{1/2}$, (C) $\lambda \propto E^{-1/2}$, (D) $\lambda \propto E^2$, (E) $\lambda \propto E^{-2}$, (F) $\lambda \propto E^{-1}$.

Answer: (C); see Eq. [22.7].

Question 22.11

Bohr used Newton's second law in the specific form

$F_{net} = m \cdot v^2/r$, not in the form $F_{net} = m \cdot a$ (a is acceleration, m is mass, v is speed, and r is radius). When is Newton's second law used in this form? (A) when objects move with constant velocity, (B) when objects move along circular paths, (C) when the total energy is conserved, (D) in cases of uniform circular motion, (E) when the system consists of two interacting objects.

Answer: (D)

Question 22.13
On which variable does the frequency of radiation emitted from a hydrogen atom depend linearly? (A) the quantum number of the initial orbit, (B) the quantum number of the final orbit, (C) the Rydberg constant, (D) the wavelength of the emitted radiation in a vacuum, (E) none of the above.

Answer: (E). The Rydberg constant is not a variable.

Question 22.15
When a hydrogen atom absorbs a photon of energy $h \cdot f$, the kinetic energy of the electron that transfers to an excited state changes by: (A) zero, (B) ½ $h \cdot f$, (C) $h \cdot f$, (D) $2 \cdot h \cdot f$, (E) $-h \cdot f$.

Answer: (E)

Question 22.17
The quantum number n can increase to infinity in Bohr's hydrogen atom. Does this mean that the possible frequencies of its spectral lines also increase without limit?

Answer: No, because the energy of the electron in the hydrogen atom is proportional to $1/n^2$.

Question 22.19
Why has an electron in the binding σ–orbital in an H_2 molecule a lower total energy than in the lowest energy state of the hydrogen atom ($n = 1$)? (A) Because it is attracted by two positive charges at the same time. (B) Because it is screened by the other electron in that orbital. (C) Because it occupies a larger space (Heisenberg's uncertainty principle). (D) Because it can no longer radiate. (E) None of the above.

Answer: (C). Answer (A) has some merit in a classical sense, but it is more complicated to show this as the two positive centres are well separated.

Question 22.21
Identify the molecule for which Fig. 22.27 shows the term scheme of the linear combination of atomic orbitals.

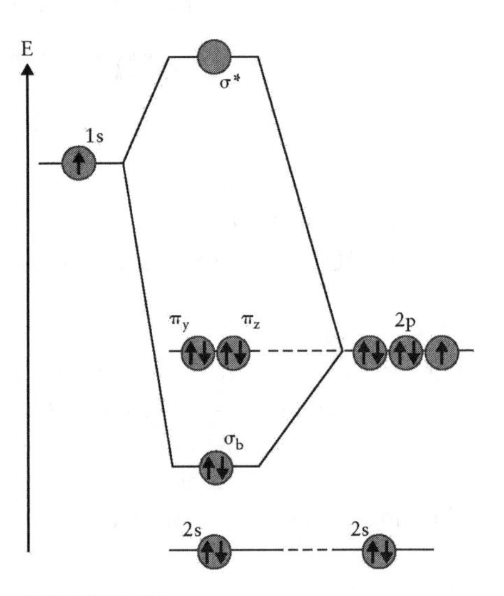

FIGURE 22.27

Answer: The easiest way to identify the molecule is to study the two atoms, shown left and right in the term scheme in Fig. 22.27. The atom shown at the left has only a single electron, thus, it is hydrogen. The atom at the right has 7 electrons in its outer shell with quantum number $n = 2$, i.e., 2 electrons in the 2s orbital and 5 electrons in the 2p orbital. We can identify this atom by consulting the periodic system: fluorine has 7 electrons in the $n = 2$ orbitals, in addition to two electrons in the 1s orbital. Thus, the molecule shown in Fig. 22.27 is hydrofluoric acid, HF.

Question 22.23
Must an atom first be ionized to emit light?

Answer: No. It can absorb energy in the ground state to transfer into any of its excited states, then emit light of the corresponding frequency.

Question 22.25
The ionization energies for Na, K, and Rb are 5.14 eV, 4.34 eV, and 4.18 eV, respectively. Why are the values decreasing in this order?

Answer: During ionization of a neutral atom the outmost valence electron is removed. This electron is in the sequence Na–K–Rb in an orbit with increasing quantum number ($n = 3$ for sodium, $n = 4$ for potassium, and $n = 5$ for rubidium).

ANALYTICAL PROBLEMS

Problem 22.1
Calculate the angular momentum of Earth as it moves around the Sun.

Solution: The magnitude of the angular momentum of an object in circular motion is given in Eq. [1] but taking $\varphi = 90^0$ because the velocity of the object is always perpendicular to the radial direction. For the Earth moving around the Sun we know that the average distance between Earth and the Sun is 1.5×10^{11} m and that the time for one orbit is a year which is equal to 3.15×10^7 s. Thus, the average speed of the Earth is 3.0×10^4 m/s.

The angular momentum of the Earth as it moves about the Sun is then $L = 2.69 \times 10^{40}$ kg · m²/s.

Problem 22.3
Fig. 22.28 shows four point–like objects that are connected by massless strings and rotate about the origin with an angular speed of 2 revolutions per second. Use $L = 1.0$ m. If we shorten their distances to the origin to 0.5 m by pulling the strings shorter, what is their new angular speed? Can you use the result to draw conclusions about the rotation of a collapsing star?

Solution: $\omega = 8.0$ revolutions/s. A collapsing star rotates faster and faster.

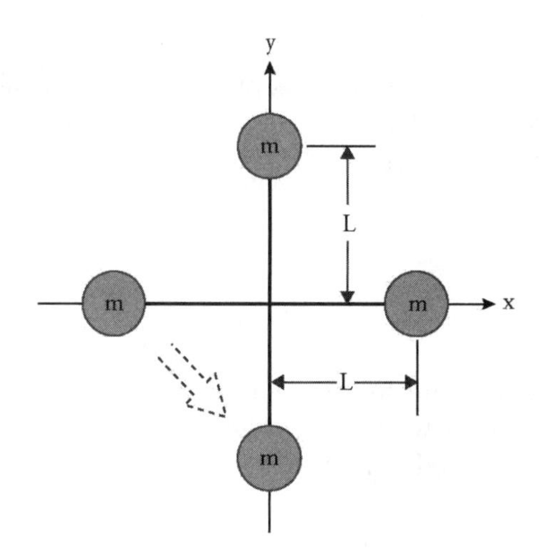

FIGURE 22.28

Problem 22.5
A hydrogen atom is in its first excited state ($n = 2$). Using Bohr's atomic model, calculate
(a) the radius of the electron's orbit,
(b) the potential energy of the electron, and
(c) the total energy of the electron.

Solution to part (a): The radius of the lowest orbit is called the Bohr radius. The radii of the excited states of the hydrogen atom are related in a simple fashion to the Bohr radius: $r(n) = n^2 \cdot r_{Bohr}$. Thus, we find for $n = 2$ a radius of $r = 2.12 \times 10^{-10}$ m = 0.212 nm.

Solution to part (b): The only form of potential energy relevant in the atomic shell is the electric potential energy. For $n = 2$ we find:

$$E_{el} = -\frac{e^2}{4 \cdot \pi \cdot \varepsilon_0 \cdot r}$$

$$= -\left(9 \times 10^9 \frac{\mathrm{N \cdot m^2}}{\mathrm{C^2}}\right) \frac{(1.6 \times 10^{-19}\ \mathrm{C})^2}{2.12 \times 10^{-10}\ \mathrm{m}} \quad \mathbf{(2)}$$

$$= -1.09 \times 10^{-18}\ \mathrm{J} = -6.8\ \mathrm{eV}$$

Solution to part (c): There are two ways to calculate the total energy. Calculating the kinetic energy and adding the kinetic energy and the result of part (b) requires a longer calculation. A shorter approach is to use Eq. [22.24]:

$$-E_{el} = 2 \cdot E_{kin} \quad \mathbf{(3)}$$

Thus, $E_{total} = -5.5 \times 10^{-19}$ J, which is equal to – 3.4 eV.

Problem 22.7
Calculate the electric force on the electron in the ground state of the hydrogen atom.

Solution: The force acting on the electron is the Coulomb force. Using for the distance between electron and proton the Bohr radius, we find:

$$|\boldsymbol{F}| = \frac{1}{4 \cdot \pi \cdot \varepsilon_0} \cdot \frac{e^2}{r_{Bohr}^2}$$

$$= \left(9 \times 10^9 \frac{\mathrm{N \cdot m^2}}{\mathrm{C^2}}\right) \frac{(1.6 \times 10^{-19}\ \mathrm{C})^2}{(5.29 \times 10^{-11}\ \mathrm{m})^2} \quad \mathbf{(4)}$$

$$= 8.2 \times 10^{-8}\ \mathrm{N}$$

Problem 22.9
Hydrogen atoms emit photon wavelengths $\lambda = 656$ nm. Which transition did the hydrogen atom undergo to emit this photon?

Solution: We convert the given wavelength into a photon energy using $E = h \cdot f$ and using $c = \lambda \cdot f$:

$$E_{trans} = \frac{h \cdot c}{\lambda} = \frac{(6.6 \times 10^{-34}\ \text{J s})\left(3 \times 10^{8}\ \frac{\text{m}}{\text{s}}\right)}{6.56 \times 10^{-7}\ \text{m}} \quad \textbf{(5)}$$

$$= 3.02 \times 10^{-19}\ \text{J} = 1.89\ \text{eV}$$

In the next step we search for a combination of hydrogen levels in Table 22.1 which corresponds to this energy. We find the transition between $n = 2$ and $n = 3$. We know further that the hydrogen atom has emitted the photon with the energy calculated in Eq. [5]. Emission corresponds to a loss of energy, thus, the transition occurred from $n = 3$ to $n = 2$, i.e., the atom was excited prior to the transition.

Table 22.1

n	E_{total} (H)	E_{total} (He)
1	– 13.6 eV	– 54.4 eV
2	– 3.4 eV	– 13.6 eV
3	– 1.51 eV	– 6.0 eV
4	– 0.85 eV	– 3.4 eV
5	– 0.54 eV	– 2.2 eV

Problem 22.11

The size of Rutherford's atom is about 0.1 nm.

(a) Calculate the speed of an electron that moves around a proton, based on their electrostatic attraction, if they are separated by 0.1 nm.

(b) Calculate the corresponding de Broglie wavelength for Rutherford's electron.

Solution to part (a): The only force that acts on the electron in orbit around the proton is Coulomb's force. This force provides the centripetal acceleration of the electron that causes it to travel in uniform circular motion around the proton. We write:

$$\frac{e^2}{4 \pi \varepsilon_0 r^2} = m_{electron} \frac{v^2}{r} \quad \textbf{(6)}$$

We re-arrange Eq. [6] for the velocity $v = 1.6 \times 10^6$ m/s.

Solution to part (b): The de Broglie wavelength is calculated from Eq. [22.4] as $\lambda = h/(m \cdot v) = 0.45$ nm.

Problem 22.13

How much energy is required to ionize hydrogen when it is in the state with $n = 3$?

Solution: When hydrogen is in the state with $n = 3$, we can find the energy it takes to be ionized by using Eq. [22.26], after substituting the value of the Rydberg constant from Eq. [22.28]. We take $n_{\text{initial}} = 3$ and $n_{\text{final}} = \infty$ because the final state is for the atom to be ionized. The energy is then $\Delta E_{\text{ionization}} = 1.51$ eV.

Problem 22.15

The intensity threshold of human dark–adapted vision at $\lambda = 500$ nm is 4.0×10^{-11} J/(m² · s). When a person's pupil is open, light enters the eye through a circular area with a diameter of 8.5 mm. How many photons reach the retina per second at the dark–adaptation threshold?

Solution: The area of the open pupil of the eye, which is circular with a radius of (8.5 mm)/2 = 4.25 mm, is $A = 1.13 \times 10^{-4}$ m². The given intensity threshold I is used as the amount of energy per unit area per second coming through the pupil. If that energy is in the form of 500 nm photons, then we can use the energy per photon to find the number of photons per second.

The energy of a single photon at $\lambda = 500$ nm is $\varepsilon = h \cdot f = 4.0 \times 10^{-19}$ J. Therefore the number of 500 nm photons per second is $N = I \cdot A/\varepsilon = 5700$ photons.

Problem 22.17

The K shell ionization energy of Cu is 8980 eV and the L shell ionization energy is 950 eV. Determine the wavelength of the K_α X–ray emission of Cu.

Solution: The K_α transition corresponds to a transfer of an electron from the L shell to a vacancy in the K shell. The associated energy of the transition is:

$$\Delta E = E_{final} - E_{initial}$$
$$= -8980\ \text{eV} - (-950\ \text{eV}) = -8030\ \text{eV} \quad \textbf{(7)}$$

which corresponds to $\Delta E = -1.28 \times 10^{-15}$ J. This value is converted to a wavelength with Planck's constant and the vacuum speed of light:

$$\lambda = \frac{c}{f} = \frac{h\,c}{|\Delta E|}$$
$$= \frac{(6.6 \times 10^{-34}\ \text{J s})\left(3 \times 10^{8}\ \frac{\text{m}}{\text{s}}\right)}{1.28 \times 10^{-15}\ \text{J}} \quad \textbf{(8)}$$
$$= 1.55 \times 10^{-10}\ \text{m} = 0.155\ \text{nm}$$

CHAPTER TWENTY–THREE

Radiation: Nuclear physics and magnetic resonance

MULTIPLE CHOICE AND CONCEPTUAL QUESTIONS

Question 23.1

A tremendous amount of energy is released per nuclear decay in the fuel cells of a nuclear power plant or in an atomic bomb explosion. Where does this energy come from?

Answer: This energy was stored as mass in the parent nuclei before the decay.

Question 23.3

Why is the binding energy of a nucleon in an atomic nucleus much larger than the chemical binding energy in a molecule?

Answer: Because the nuclear force is a very strong force. We can estimate how strong it is from the fact that it is larger than the electric repulsive force between two protons that are separated by a distance of the order of the size of a nucleus.

Question 23.5

Is the radioactive decay law linear? If not, can you rewrite it such that a plot yields a straight line?

Answer: The radioactive decay law is not linear, see Eq. [23.13]:

$$N(t) = N_0 \, e^{-\lambda \cdot t} \quad (1)$$

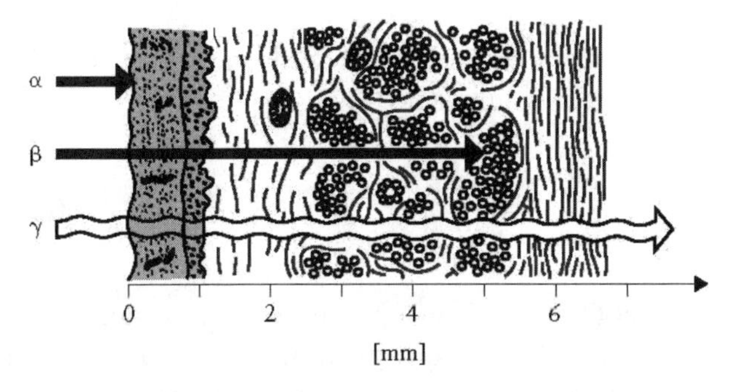

FIGURE 23.6

It leads to a straight line in a logarithmic plot.

Question 23.7

Fig. 23.6 and Table 23.1 indicate that alpha particles penetrate the skin only to a shallow depth. Why, then, is this type of radiation considered particularly dangerous, as indicated by its large radiation factor in Table 23.2?

Table 23.1

Radiation type	Energy	Range
α–particles	5 MeV	40 μm
β–radiation	20 keV	10 μm
β–radiation	1 MeV	7 mm
γ–radiation	20 keV	6.4 cm
γ–radiation	1 MeV	65 cm
neutrons	1 MeV	20 cm

Table 23.2

Radiation type	w_R
X–ray and γ–radiation	1
β^+, β^-	1
n^0	5 – 10
α	10

Answer: The α–particle causes much more ionization along its track than β–particles or γ–rays. Thus, it causes greater damage to tissue. Its effect is still less superficial than UV–radiation. Also, α–emitters can be swallowed or inhaled and then damage internal organs.

Question 23.9

You work in a lab that uses radioactive materials. A spill occurs due to an accident. What is your first reaction?

Answer: Increase your distance d to the spill as fast as possible because the radiation damage is proportional to $1/d^2$.

Question 23.11
The activity of an unknown radioactive isotope reduces to 96% of the original value in 120 minutes. What is its half–life? Choose the closest value. (A) 610 min, (B) 2040 min, (C) 2640 min, (D) 4120 min.

Answer: (B)

Question 23.13
The emitted particle in beta decay has a wide range of possible energies, but the energy of the helium nucleus emitted in alpha decay has a decay–specific value. What explains this difference?

Answer: In beta decay, additional neutrinos (or anti–neutrinos) are released that carry a variable fraction of the released energy.

Question 23.15
Why are X–ray cassettes in hospitals not affected when an alpha emitter is in close proximity, but deteriorate when a beta–emitter comes close?

Answer: Alpha particles cannot penetrate the thin metal sheet of which the cassette is made while beta particles can penetrate it and expose the film.

Question 23.17
In this chapter, we discussed the tidal waves and their effect on Earth's rotation. Given that the Sun is much bigger than the Moon, shouldn't the effect due to the Sun be larger than that due to the Moon?

Answer: Tidal waves result because the gravitational pull of the Moon on surface–water occurs at a closer distance than the gravitational pull on the solid Earth (which acts on its centre). Two tidal waves result because the Moon attracts the ocean water in closer proximity, and pulls the Earth away below the ocean water on the opposite side. Thus, the height of a tidal wave is a function of the relative difference of the distance from the attracting astronomical object to the surface and to the centre of Earth. Even though the Sun is much more massive, its this relative difference in distance is much smaller in this case. The Sun still does enhance tidal waves, which are then called *spring tide*.

ANALYTICAL PROBLEMS

Problem 23.1
The nucleus of the deuterium atom consists of one proton and one neutron. What is the binding energy of this nucleus if the mass of the deuterium nucleus is given as $2.014102 \cdot u$?

Solution: A proton has a mass of $1.007276 \cdot u$, a neutron $1.008665 \cdot u$ with $1\ u = 1.6605677 \times 10^{-27}$ kg. The binding energy of the deuteron is provided by the difference between the deuteron nucleus mass and the combined mass of its constituents, i.e., one proton and one neutron:

$$\begin{aligned} \Delta m &= m_D - m_{p^+} - m_{n^0} \\ &= 2.014102 \cdot u - 1.007276 \cdot u - 1.008665 \cdot u \quad \textbf{(2)} \\ &= -1.839 \times 10^{-3} \cdot u \end{aligned}$$

Using Einstein's formula we convert the loss of mass into an energy:

$$\begin{aligned} \Delta E &= \Delta m \cdot c^2 = 2.388 \times 10^{-3}\, u \left(3 \times 10^8\ \frac{\text{m}}{\text{s}} \right)^2 \quad \textbf{(3)} \\ &= 2.75 \times 10^{-13}\ \text{J} = 1.72\ \text{MeV} \end{aligned}$$

In this calculation we established the mass to energy conversion $1\ u = 931$ MeV. Comparing the result in Eq. [3] with the result for the helium nucleus in Example 23.1, where we found $E_{\text{binding}}(\text{He}) = 28.3$ MeV, shows that the deuteron is weakly bound. Indeed, the largest binding energies (per nucleon) occur for the most stable nuclei which are located near iron in the periodic system.

Problem 23.3
Nuclear waste from power plants may contain ^{239}Pu, a plutonium isotope with a half–life of 24000 years. How long does it take for the stored waste to decay to 10% of its current activity level?

Solution: We calculate the decay constant from the half–life of the plutonium isotope:

$$\lambda = \frac{\ln 2}{T_{1/2}} = \frac{\ln 2}{24000\ \text{yrs}} = 2.89 \times 10^{-5}\ \text{yrs}^{-1} \quad \textbf{(4)}$$

Note that the value is given in unit 1/yrs. Although it is always advisable to convert to SI units (here 1/s), we will find that it is not necessary for the current problem.

Now we apply Eq. [1] with $N(t) = 0.1 \cdot N_0$ which corresponds to 10% of the initial amount of radioactive plutonium. This leads to:

$$t = -\frac{\ln(0.1)}{2.89 \times 10^{-5}\ \text{yrs}^{-1}} = 79700\ \text{yrs} \qquad \textbf{(5)}$$

Just waiting isn't good enough when it comes to dealing with nuclear waste!

Problem 23.5
A tracer study drug contains 11 kBq of a technetium isotope, ^{99}Tc, which has a half–life of $T_{1/2}$ = 363 minutes. Technetium can be used as a substitute for ^{131}I in tracer studies of the thyroid gland. What is the activity of the drug when it is used after 3 hours?

Solution: We use the time dependence of the activity (in unit Bq):

$$\begin{aligned} A(t_{final}) &= A(t = 0)\, e^{-\lambda \cdot t_{final}} \\ &= A(t = 0) \exp\left\{-\frac{ln2 \cdot t_{final}}{T_{1/2}}\right\} \end{aligned} \qquad \textbf{(6)}$$

We substitute the given values, and use 3 h = 10800 s and 363 min = 21780 s:

$$\begin{aligned} \frac{\Delta N}{\Delta t} &= \left(\frac{\Delta N}{\Delta t}\right)_0 \exp\left(-\frac{ln2 \cdot t}{T_{1/2}}\right) \\ &= (1.1 \times 10^4\ \text{s}^{-1}) \exp\left(-\frac{\ln 2 \cdot 10800\ \text{s}}{21780\ \text{s}}\right) \end{aligned} \qquad \textbf{(7)}$$

The activity after 3 hours results as $\Delta N/\Delta t$ = 7800 Bq = 7.8 kBq.

Problem 23.7
An archeological wood sample is analysed. What is its age if its ^{14}C content is 15% of a living wood sample?

Solution: If the ^{14}C content is only 15% that of a living wood sample, then $N/N_0 = 0.15$. We can use this along with the half–life of ^{14}C (5730 years) to find the age of the wood. Note that in the calculation it is not necessary to convert the half–life into seconds. As long as we are careful with keeping track of units then the final answer will come out in units of years as well. We start with Eq. [4]: $\lambda = 1.21 \times 10^{-4}\ \text{yrs}^{-1}$. Eq. [1] is then solved for t = 15700 yrs.

Problem 23.9
A radioactive sample contains 5.0 μg of pure ^{15}O with a half–life of 2.03 min.
(a) How many radioactive nuclei does the sample contain?
(b) What is the activity of the sample after 5 hours?

Solution to part (a): Pure ^{15}O has a molar mass of M = 15 g/mol. Therefore, the number of atoms in the given sample will be $N_0 = 2.0 \times 10^{17}$ atoms.

Solution to part (b): The half–life of this isotope is given as 2.03 min = 122 s so we can calculate the decay constant as $\lambda = 5.68 \times 10^{-3}\ \text{s}^{-1}$.

The initial activity is $A_0 = 1.14 \times 10^{15}$ Bq, and the activity after 5 hours (18000 s) is then found from Eq. [6]: $A_{final} = 4.5 \times 10^{-30}$ Bq, i.e., $A_{final} = 0$: Note that 5 hours is a very long time compared to the half–life of 2.03 min, so there is almost no activity after 5 hours.

Problem 23.11
How much heavy ion radiation does the same damage as 10 Sv X–rays?

Solution: If the equivalent dose $D_{equivalent}$ is 10 Sv, this is the same whether one is talking about X–rays or heavy ions (α–particles). The difference comes from the radiation factor, w_R, which refers to the biological impact of the radiation. We can see this in Eq. [23.22].

Since the radiation factor is 1 for X–rays and 10 for α–particles, and assuming the radiation affects the whole body with a tissue factor w_T = 1, then we find $D_{absorbed}$ = 10 Gy for X–rays buy only 1 Gy for α–particles.

Problem 23.13
A standard man is exposed to a whole–body radiation of 0.25 Gy. What amount of energy (in unit J) is added to the person's body?

Solution: The value of the absorbed dose of radiation is units of grays (Gy) is defined as the radiation energy absorbed per unit mass. For a whole–body radiation, the value of 0.25 Gy must be multiplied with the mass of the person, E = 17.5 J.

Problem 23.15
A sample of 2.0 Gy is administered during a cancer treatment. If the entire radiation is absorbed by a tumour of mass m = 250 g,
(a) how much energy does it absorb, and

(b) how much does its temperature increase? *Hint*: Use thermodynamic data for water to simulate tissue.

Solution to part (a): The energy dose $D_{absorbed}$ in unit gray (Gy) is the measure of energy E absorbed per unit mass. Therefore, $D_{absorbed} = E/m$, so $E = D_{absorbed} \cdot m = 0.5$ J.

Solution to part (b): In absorbing this energy, the temperature of the tumour will increase. See Chapter 6 for a discussion of thermal energy. We will approximate the thermodynamic properties of the tumour by those of water, taking the specific heat capacity of water from Table 6.4. From the definition of heat in Eq. [6.90], the change in temperature will be $\Delta T = Q/(c \cdot m) = 480$ μK.

Note in this solution that the heat was the energy added to the tumour by the cancer treatment.

Problem 23.17

A conductor forms a single circular loop of 0.5 m radius. It carries a current of $I = 2.0$ A and is located in a uniform magnetic field of $B = 0.4$ T.

(a) What is the maximum torque that acts on the loop?

(b) What is the angle between the magnetic field and the plane of the loop when the torque is one–half of the value found in part (a)?

Solution to part (a): To calculate the torque on a current loop in a magnetic field we need the current, the field, the area of the loop and the relative orientation of the loop and the field direction. Here, the current and the magnetic field are given. From the geometry of the loop we can find its area. Finally, the relative orientation of the loop and the field direction may be chosen so that the torque is a maximum; i.e. so that the magnetic field direction and the normal vector of the area enclosed by the loop are perpendicular.

The current loop in this problem has the shape of a circle, so we can find its area from its radius as $A = 0.79\ m^2$.

We can now calculate the torque on the loop using Eq. [23.42]: $\tau_{max} = 0.63$ N · m. Note that we have chosen the value of $\varphi = 90^0$ in Eq. [23.42] so that the torque is maximized.

Solution to part (b): The torque will be one–half the value of part (a) when the angle between the loop and the magnetic field direction has changed. Note that the current, area of the loop and external field remain constant so that the torque will be one–half its maximum value when: $\sin\varphi = 0.5$ or $\varphi = 30^0$.

This is the angle between the magnetic field direction and the normal vector of the area enclosed by the current loop. The angle between the magnetic field and the *plane* of the loop is $90^0 - 30^0 = 60^0$ since the normal vector is perpendicular to the plane of the loop.